Eat
Your
Vegetables!

Eat Your Vegetables!

More than 200 irresistible recipes from
the authors of DON'T TELL 'EM IT'S GOOD FOR 'EM

NANCY BAGGETT · RUTH GLICK
and GLORIA KAUFER GREENE

Times
BOOKS

Library of Congress Cataloging in Publication Data

Baggett, Nancy, 1943–
Eat your vegetables!

Includes index.
1. Cookery (Vegetables) I. Glick, Ruth, 1942–
II. Greene, Gloria Kaufer, 1950–   III. Title.
TX801.B26 1985   641.6'5   85-40276
ISBN 0-8129-1201-2

Illustrated by: Meg Galub

Coordinating Editor: Rosalyn T. Badalamenti

Designed by Giorgetta Bell McRee/Early Birds

Manufactured in the United States of America
9 8 7 6 5 4 3 2
First Edition

To all mothers who have ever said,
"Eat your vegetables!"

Acknowledgments

We especially want to thank Kathleen Moloney, senior editor at Times Books, for suggesting this book and for her enthusiastic support throughout the project.

Many thanks also go to coordinating editor Rosalyn T. Badalamenti, for her professional expertise and careful attention to detail.

We would also like to acknowledge the United States Department of Agriculture and the United Fresh Fruit and Vegetable Association for their research assistance.

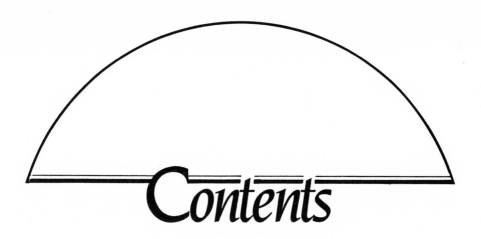

Contents

Introduction

Modern cooks are making an exciting discovery. Vegetables can be the life of the culinary party.

They add gorgeous yet absolutely natural colors—the brilliant orange of sweet potatoes; the emerald green of lightly steamed broccoli.

They provide satisfying texture—the snappy crunch of celery; the velvety smoothness of Boston lettuce.

They're packaged in an amazing variety of shapes and sizes—the petal-like leaves of an artichoke; the perfect miniature roundness of a single green pea.

They open up a world of contrasting tastes—the succulent sweetness of just-picked corn, the zesty bite of Belgian endive; the richness of vine-ripened tomatoes.

And, as a bonus, these gifts of the earth are bursting with the vitamins and minerals we all need for good health and long life.

Yet when we were growing up, we weren't aware of all the wealth that vegetables had to offer. Usually, they were just afterthoughts, tacked on to round out a meal, and more often than not they were canned or frozen. There was no special thought given to preparation. They were simply dumped into a pot and boiled. Fresh vegetables did make it into the salad bowl, but the combination was almost always the same: iceberg lettuce, celery, carrots, and tomatoes. And we didn't even know what we were missing!

This wasn't really anybody's fault. It was the era when convenience foods were an exciting novelty. Then, too, fresh vegetables just weren't as available as they are to-day.

Thank goodness things have changed. Improved growing techniques and advances in transportation have put a bonanza of wonderful fresh vegetables in supermarkets year round. Crisp, snowy heads of cauliflower, vibrant sweet red peppers, satiny purple eggplants, and dozens of other varieties tempt us at every turn.

This exciting abundance has made it possible—and enticing—to broaden our culinary horizons. Vegetables have sparked interest in exploring regional and foreign cuisines, and in making dining a memorable experience. What's more, there's an increasing awareness that eating fresh vegetables can have a positive effect on health.

Modern cooks are naturally interested in taking advantage of all this. However, when they turn to classic dishes, they find that many unfortunately sabotage the goodness of vegetables. Some recipes mask the fine natural flavors with heavy sauces and load down vegetables with unhealthful, calorie-laden fats. Others destroy color, texture, and nutrients by overcooking.

We accepted the challenge to do better when we decided to write *Eat Your Vegetables!* And the results have been enormously satisfying in every sense of the word. Using the experience gained from our previous cookbook, *Don't Tell 'Em It's Good for 'Em*, we found we could devise wonderfully tasty recipes without relying heavily on butter, cream, eggs, and cheese, as many other cookbooks do.

Our recipes are enlivened with herbs, spices, and other "seasonings" like wine, lemon juice, and rich vegetable and meat broths. They make use of light, appealing sauces and dressings that include only modest amounts of butter, cream, oil, or mayonnaise. They combine vegetables that accent one another's flavor and color. They take advantage of easy cooking techniques which accentuate the best qualities each vegetable has to offer. And some team up vegetables with meat to enhance the good taste of both. (However, we haven't included any cured or smoked meats because research has shown that they are potentially carcinogenic.)

If you thumb through this cookbook, you'll immediately see that the chapters are organized alphabetically by vegetable. Throughout, we have emphasized fresh produce. However, when frozen or canned vegetables are convenient and tasty substitutes for fresh, we don't hesitate to call for them.

Part of our aim has been to introduce you to a wide variety of vegetables that you may have been hesitant about trying in the past. Suppose you'd like to "get your feet wet" with turnips but can't quite imagine sitting down to a bowlful of them. We've included some recipes that will help you ease them into your culinary repertoire. For example, our Peasant Soup is a savory blend of turnips and other more popular vegetables. And unless you advertise it, no one will know that turnips contribute the zesty flavor to our Turnip and Carrot Slaw.

Indeed, many of our recipes combine less familiar vegetables with ones that are more widely eaten. We've teamed up spinach and kale in our elegant Spinach Pie. Our Turkish-Style Mixed Vegetables include okra and eggplant along with tomatoes, zucchini, and green peppers.

We've also used grated or finely chopped vegetables where you might not expect them. Our Garden Meat Loaf, for example, capitalizes on the moistness and flavor of carrots and celery. And our Pineapple-Carrot Gelatin Mold incorporates both carrots and cabbage. Grated, chopped, and puréed vegetables also enhance many of our soups, stews, and sauces. (You may find a food processor or blender handy for preparing these.)

Many of our recipes combine several vegetables. So when you're looking for ways to use a particular one, don't just check the chapter spotlighting it. Also refer to the index to find where else in the book the ingredient appears.

You'll also notice that some chapters are longer than others. That's because all vegetables are not created equal! We've chosen to showcase the most healthful ones—those packed with vital nutrients.

We have lots of recipes featuring carrots, sweet potatoes, rutabagas, and winter squash because these vegetables are very high in beta-carotene, which may help protect the body against several types of cancer.

Also highlighted are cabbage, broccoli, cauliflower, Brussels sprouts, collards, kale, turnips, and other members of the cruciferous (cabbage) family. Research suggests that they actually stimulate the body to produce an anti-cancer enzyme.

Many of these, and other vegetables, such as peppers, tomatoes, bean sprouts, spinach, and scallions, are also high in vitamin C, which may help prevent cancer of the digestive tract by blocking the formation of highly carcinogenic chemicals called nitrosamines.

(Nutritional data for every vegetable is included in its chapter introduction.)

As a bonus, *Eat Your Vegetables!* offers more than just recipes. Each chapter begins with some interesting background, followed by practical information. You'll discover when each vegetable is available, how to choose the best of each variety, and how to store it. In addition, you'll learn basic preparation and cooking techniques—along with some simple serving suggestions that will add interest and taste appeal quickly and easily.

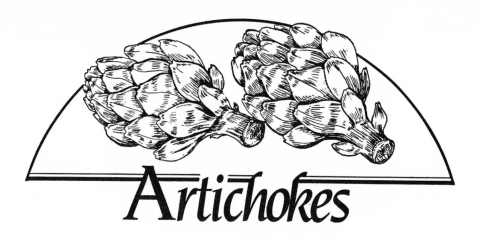

Artichokes

A native of the Mediterranean region, the artichoke is actually a large, thistle-like bud on the end of a long, thick stem. If an artichoke is not harvested, it eventually blossoms into a composite cluster of attractive, violet-blue flowers.

Virtually all the artichokes eaten in this country are dark green "globe" artichokes, so named because of their basically globular shape. In provincial France, spindle-shaped, purplish artichokes are sometimes preferred.

During the Renaissance in Europe, artichokes became quite fashionable, particularly among monarchs and courtiers. Catherine de Médicis—who may have brought artichokes from Italy to France—is said to have eaten so many at a sixteenth-century feast that she "nearly burst." And this was in spite of the fact that artichokes were then considered to be aphrodisiacs, and, thus, unsuitable food for respectable women!

To this day, the artichoke is quite popular in both France and Italy, where it is relatively inexpensive and commonplace. In the United States, however, it is still generally considered to be a "gourmet" vegetable—unfortunately, with a price to match.

Most of our artichokes are grown in the vicinity of Castroville, California, a small coastal community south of San Francisco. This "artichoke capital of the world," as the city calls itself, has a cool, foggy, relatively frost-free climate that is ideal for cultivation of the vegetable.

There, one can purchase delectable fresh artichoke "hearts." These come from tiny, young artichokes which have not fully developed the coarse, fuzzy choke of completely mature specimens, and are thus entirely edible except for some tough outer leaves and the stems.

An artichoke "bottom" (often mistakenly called the "heart") is the fleshy concave disk that remains when the stem, leaves, and choke are removed from a mature, fully grown artichoke. The bottom, which is about 2 to 3 inches in diameter and about ½ to ¾ inch thick, is considered the choicest part of the mature vegetable.

In large, mature artichokes, only the bottom, a small bit of flesh on each leaf, and the inside of the stem are edible. Nonetheless, whole artichokes provide low-calorie finger food that is not only tasty, but actually fun to eat.

Availability: Generally, whole fresh mature globe artichokes are available throughout the year, with the peak season in April and May, and lowest supply in summer. Both the hearts and bottoms are available canned, and the hearts sometimes come frozen. *Fresh* artichoke hearts are not commonly available outside California.

Choosing the Best: When buying fresh artichokes, look for uniformly solid heads that are heavy for their size and have compact, fleshy leaves that are firm, not wilted or shriveled. As flavor is not affected by size, and artichokes are usually priced per item, choose the largest ones.

When artichokes have been exposed to frost during growth, the leaves become brownish and blistered. Though a bit unattractive, this "winter kist" condition, as it is called by the growers, does not affect quality. In fact, some experts say it may even improve the flavor.

Nutritional Value: Artichokes are high in non-nutritive fiber, phosphorus, and potassium. Their caloric content goes up with storage and maturity, dramatically increasing from about 20 to 70 calories per artichoke (and also varying with size).

Storage: To keep artichokes from drying out, refrigerate them in plastic bags or covered containers. They are best if used within 2 to 4 days of purchase.

Preparation and Basic Cooking: Trim each artichoke as follows, discarding all trimmings: Use a sharp knife to cut off the stem even with the base (so the artichoke can rest upright without falling over). Cut off about ¾ inch of the top of the leaf cone. Break off (do not cut) the row of small leaves around the base of the artichoke. For a classic presentation, use scissors or kitchen shears to clip off the thorny tip of each leaf. If the artichokes are not to be cooked immediately, coat the trimmed edges with a little lemon juice so they do not turn brown.

To cook the artichokes, sit them in a pot just large enough to hold them side by side, supporting one another. Add enough water to come about halfway up the sides of the artichokes. Cover the pot and bring the water to a boil over high heat. Lower the heat and simmer the artichokes, covered, for about 40 to 45 minutes, or until an outside leaf pulls out easily, and the flesh on the leaf is tender. Use tongs or a slotted spoon to carefully remove the artichokes from the water and drain them well before serving.

Simple Serving Suggestions: Hot, whole artichokes are quite delicious served simply with warm melted butter or hollandaise sauce. Pluck each leaf from the artichoke, dip it into the butter or sauce, and then pull it through your teeth to scrape off the meaty flesh on the inside bottom of each leaf. Continue with all the leaves. (Some thin, whitish, inner leaves may be almost entirely edible.)

When you get down to the small white cone of thorny, purple-edged leaves in the center, use a spoon to carefully remove and discard it along with the hairy choke underneath, leaving the bottom intact. Cut the bottom into bite-sized pieces, and enjoy it with more butter or sauce. Whole cold artichokes may be similarly served with a vinaigrette. As an appetizer or vegetable side dish, one whole artichoke serves 1 to 2 persons.

For a very elegant presentation, serve each cooked and trimmed artichoke bottom topped with a poached egg and hollandaise sauce. This is perfect for a brunch main course or a first course at a dinner party.

MARINATED ARTICHOKE HEARTS

(salad or appetizer)

1   12-ounce (approximately) can water-packed quartered artichoke hearts
¼   cup vegetable oil
2   tablespoons apple cider vinegar
¾   teaspoon sugar
¼   teaspoon dried marjoram leaves
¼   teaspoon dried basil leaves
¼   teaspoon celery salt
⅛   teaspoon dried oregano leaves
    Pinch of onion powder
    Pinch of black pepper, preferably freshly ground

Drain the artichoke hearts well in a colander. In a small bowl or cup, combine all the remaining ingredients. Stir to mix well.

Put the drained artichoke hearts into a bowl and pour the dressing over them. Toss to mix well. Cover and refrigerate for several hours so that the flavors mingle. Toss the salad again before serving.

Makes 6 to 8 servings.

ARTICHOKE, AVOCADO, AND GRAPEFRUIT SALAD

Dressing
½ cup vegetable oil
⅓ cup red wine vinegar or apple cider vinegar
1 teaspoon sugar
  Scant ½ teaspoon powdered mustard
½ teaspoon celery salt
⅛ teaspoon dried basil leaves
⅛ teaspoon black pepper, preferably freshly ground

Salad
1 small head escarole, separated into leaves, washed, and well drained
2 large pink grapefruits, peeled and segmented
2 small ripe avocados, pitted and cut lengthwise into ¼-inch-thick slices
1 11½-ounce jar or can water-packed quartered artichoke hearts, well drained
  Several thin slices red onion, separated into rings (optional)

To prepare the dressing, combine all the ingredients in a cruet or a jar with a tight-fitting lid. Shake the mixture vigorously until well mixed. Refrigerate for at least 30 minutes to allow the flavors to blend.

To assemble the salad, pat the escarole dry on paper towels. Tear any very large leaves in half. Attractively arrange the leaves on a large round serving platter or in a large shallow salad bowl. Alternating the grapefruit segments and avocado slices, and working from the outside of the platter inward, arrange them in concentric circles on the escarole bed. Mound the quartered artichoke hearts in the center of the platter. If desired, garnish the platter with onion rings. The salad may now be covered and refrigerated for up to an hour or served immediately. Shake the dressing and drizzle it over the salad just before serving.

Makes 6 to 8 servings.

CHICKEN WITH ARTICHOKES AND ONIONS

(main dish)

This is a delicious and attractive dish that makes a nice main course for company. For an elegant touch, substitute whole shallots for the tiny white onions.

2   tablespoons butter or margarine
1   tablespoon vegetable oil
1   pint very small white onions (or 4 small yellow onions, cut into quarters)*
    About 2½ pounds meaty chicken pieces (skinned, if desired)
1   10-ounce package frozen artichoke hearts, thawed and drained
2   tablespoons water
2   tablespoons finely chopped fresh parsley leaves
1   teaspoon dried chervil leaves or dried marjoram leaves
2   bay leaves
¼   teaspoon salt
⅛   teaspoon black pepper, preferably freshly ground

In a large deep skillet over medium-high heat, heat the butter with the oil. Sauté the onions until they are lightly browned on all sides. Add the chicken and lightly brown the pieces, turning as necessary. Add the artichoke hearts, evenly distributing them among the chicken pieces, and the water. Sprinkle the herbs and seasonings on top of the chicken and vegetables and cover the skillet tightly. Turn the heat to low and cook the chicken and vegetables slowly, basting them often with the juices in the bottom of the skillet. Cook for about 40 minutes, or until the chicken is quite tender. (If the chicken and vegetables appear to be drying out, add a few more tablespoons of water to the skillet.) Serve with any remaining pan juices poured on top.

Makes 4 to 5 servings.

*See the chapter on Onions, page 172, for directions on how to peel small white onions.

WHOLE ARTICHOKES STUFFED WITH SALMON

(main dish)

When fresh artichokes are in season, this impressive dish is a tasty, low-calorie way to serve them. Though the preparation does take some time, it is not difficult and is well worth the effort.

4 to 5	large whole fresh artichokes
	Lemon juice, preferably fresh
1	16-ounce can pink or red salmon, drained
½	cup finely chopped fresh parsley leaves
¼	cup grated carrot
2	tablespoons plain lowfat yogurt
1	tablespoon mayonnaise
2	tablespoons fine, dry bread crumbs
1	tablespoon instant minced onions
½	teaspoon dried dillweed
⅛	teaspoon black pepper, preferably freshly ground
⅛	teaspoon onion powder
1	egg white
4	tablespoons grated Parmesan cheese, divided
4 to 5	teaspoons butter or margarine, melted

Trim each artichoke as follows (discard all trimmings): Use a sharp knife to cut off the stem even with the base so the artichoke rests upright without falling over. Cut off about ¾ inch of the top of the leaf cone. Break off (do not cut) the row of small leaves around the base of the artichoke. Use kitchen shears or scissors to clip off the thorny tip of each remaining leaf. Coat the trimmed edges of each artichoke with a little lemon juice so they do not turn brown.

To cook the artichokes, sit them in a pot just large enough to hold them side by side, supporting one another. Add enough water to come about halfway up the sides of the artichokes. Cover the pot and bring the water to a boil over high heat. Lower the heat and simmer the artichokes, covered, for about 35 minutes, or until they are almost tender. (Note: Whole artichokes are usually cooked 40 to 45 minutes, or until they are very tender and an outside leaf pulls out easily. However, in this case the artichokes will fall apart while being stuffed if they are cooked until they are completely tender.)

While the artichokes are cooking, prepare the stuffing. Mix together all the remaining ingredients *except* 2 tablespoons of the Parmesan cheese and the melted butter.

When the artichokes are ready, drain them well and let them cool until they can be handled. Gently spread apart the leaves in the center of each artichoke, and use a small spoon (a serrated grapefruit spoon works very well) to carefully scoop out the tiny, undeveloped, purple-tipped leaves and inedible hairy choke, but not the tasty artichoke bottom.

Spoon about ¼ to ⅓ cup of the stuffing into the center of each artichoke. Place the artichokes upright and close together in a casserole dish. Sprinkle the tops with the reserved Parmesan cheese and the melted butter. Pour about ½ cup water in the bottom of the casserole and bake the artichokes, uncovered, in a preheated 375-degree oven for about 20 minutes, or until the stuffing is completely heated through and the artichokes are very tender.

Serve 1 artichoke per person and provide an empty bowl for discarded leaves. To eat the artichoke, pull off an outer leaf. Hold it by the tip and pull it through your teeth, scraping off the soft flesh on the inside of the leaf. Continue until all the leaves have been sampled. Use a fork to eat the stuffing and entirely edible artichoke bottom.

Makes 4 to 5 servings.

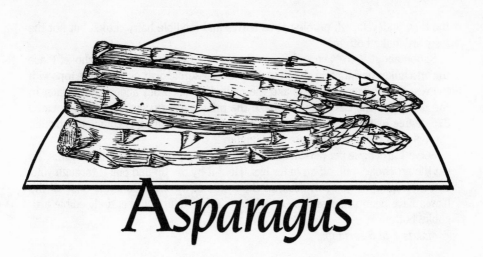

Asparagus

The first appearance of fresh asparagus in home gardens, at farmers' markets, and at produce sections of supermarkets is a sure sign of spring—long awaited by those who treasure the vegetable's wonderful taste.

Though the origin of asparagus is unknown, the spears have been cultivated—and greatly appreciated—for at least two millennia. Ancient Egyptians, Romans, and Greeks apparently ate them in a variety of different dishes. Not only did they enjoy the vegetable for its rich flavor and texture, but its shape led to the obvious conclusion that asparagus must also be an aphrodisiac.

Over the centuries, asparagus has had many enthusiastic fans, but few as devoted as the Frenchman of yore who begrudgingly offered to share his portion of the vegetable with an unexpected guest. The host instructed his cook to prepare half the asparagus with a favorite oil dressing and the rest in a different sauce his guest preferred. When, shortly before dinner, the guest keeled over from an attack of apoplexy, the Frenchman immediately ran to the kitchen and shouted, "Serve all the asparagus with oil!"

American pioneers traveling westward carried asparagus plants with them; they called the vegetable by an old name, "sparrowgrass." Interestingly, it is still called "grass" by some of those in the produce trade.

Asparagus is a member of the lily family, which also includes onions, garlic, and leeks. It grows wild all over the world, and has probably been spread not only by man but by birds, who carry the seeds from place to place. The wild varieties of asparagus look and taste very much like the cultivated ones.

Though it takes about 3 years for the perennial plants to initially become productive, they may subsequently yield seasonal crops for up to 35 years. Asparagus spears grow very quickly, and the plants can reach heights of 4 to 6 feet with vast foliage if left uncut. However, the spears are usually picked when they are only 8 to 12 inches tall.

Just about all the fresh asparagus eaten in this country has long, slender green spears. However, Europeans prize a rather different, white to yellowish-white asparagus, which has thick, fleshy spears that are grown almost entirely underground to keep them from developing chlorophyll. In the United States, this type is usually only available in cans or jars.

Fresh asparagus tastes best when cooked briefly, a fact obviously appreciated by the ancient Roman emperor Augustus, who is said to have described a task quickly done as taking "less time than to cook asparagus."

Availability: Fresh asparagus is sometimes available from late February to July. However, it is most likely to be found during April through June, the period of greatest supply. Frozen asparagus is generally in supermarkets throughout the year. Though canned asparagus is also available year round, its flavor and texture are quite different from fresh, and it is generally considered to be inferior.

Choosing the Best: Asparagus spears should be bright green, perfectly straight, firm, and brittle. The tips should be tight, compact, and pointed. Any spears with open, bent, or yellowing tips are past their prime. If possible, choose spears with less than an inch of tough, white woody part at the bottom of the stalk, as this section will have to be discarded. Avoid those spears that look shriveled or dried out. For even cooking, choose equally sized spears that are about ½ inch in diameter. Stalks that are very thin or very thick will tend to be tough and stringy.

Nutritional Value: Asparagus is high in vitamin A and has fair amounts of vitamin C and iron. It is low in calories and sodium.

Storage: Like many other seasonal vegetables, asparagus tastes best when just picked. However, if properly refrigerated, it will stay fresh for 3 to 4 days. For best results, wrap the bottoms of the stalks in damp paper towels; then place the stalks in a plastic bag to retain humidity. Or stand the spears in about an inch of water and cover them loosely with a plastic bag.

Preparation and Basic Cooking: Rinse asparagus well under cool running water, making sure to remove any sand that may be caught under the scales on the sides of the spears. Break off and discard the tough white part at the bottom of each spear. If desired, the skin may be thinly peeled from the lower half of each spear; however, this is really not necessary, particularly with very fresh asparagus spears that are not too thick.

The easiest way to cook asparagus is in a large skillet or pot of boiling water. Bring the water to a boil, add the asparagus with the spears all facing the same way (for easier serving), and simmer, uncovered, just until they are tender and bright green but not limp, about 7 to 10 minutes, depending on the thickness. Remove the spears with tongs or a slotted spoon.

Asparagus may also be steamed in a tall narrow container, such as a coffeepot. Stand the spears in about 1 to 2 inches of water and bring the water to a boil. Cover and steam the spears until tender, as above, about 8 to 12 minutes.

Another cooking method that works well is stir-frying. This produces crunchy pieces of asparagus that taste quite different from the boiled or steamed versions. Clean the spears as described above and drain them well on paper towels; then cut them diagonally into 2-inch-long pieces. (The diagonal cut gives more surface area for quick cooking.) In a wok or large skillet over medium-high heat, heat a few tablespoons of peanut or vegetable oil and a dash of salt until quite hot. Add the asparagus pieces and stir constantly until they are bright green and crisp-tender with a bit of crunch, about 3 minutes.

Simple Serving Suggestions: Boiled or steamed asparagus tastes quite delectable with a bit of butter and/or lemon juice or butter and grated Parmesan cheese. It is also good chilled and served in salads. Stir-fried asparagus pieces are not only a tasty side dish but can even be served as an hors d'oeuvre when accompanied with toothpicks.

CREAMY ASPARAGUS SOUP

Even though this soup doesn't contain a drop of cream, it still tastes rich and satisfying. In fact, one of our children who claims he hates asparagus loves this "green vegetable soup" when the telltale garnish of asparagus tips is not put on his portion.

	About 16 to 18 ounces fresh asparagus spears, approximately ½ inch thick
2	large or 3 medium-sized romaine lettuce leaves (including ribs), chopped (or substitute 2 celery stalks, chopped)
¾	cup water (plus extra for the asparagus tips)
2½	tablespoons butter or margarine
3	tablespoons enriched all-purpose or unbleached white flour
3	cups chicken broth or bouillon (reconstituted from cubes or granules)
⅛	teaspoon salt (optional)
	Scant ⅛ teaspoon black or white pepper, preferably freshly ground
⅛	teaspoon ground nutmeg
½	cup instant nonfat dry milk powder, dissolved in ¼ cup water

Wash the asparagus spears very well and drain them. Gently break off and discard the tough white part at the bottom of each stem. Cut off all the tips and set them aside. Coarsely slice the remaining part of the spears into 1-inch-long pieces. There should be about 2 cups of pieces. Put the asparagus pieces (not the tips), lettuce, and ¾ cup water into a medium-sized saucepan. Over high heat, bring to a boil; then lower the heat, and simmer, covered, for about 5 to 8 minutes, or until the asparagus pieces are just tender. Put the entire contents of the saucepan in a blender or food processor and process until puréed. Set aside.

Dry the saucepan; then melt the butter in it over medium-high heat. Add the flour and cook, stirring, for 1 minute. Stir in the broth, salt, pepper, and nutmeg. Heat, stirring, until the mixture thickens slightly and boils gently for 1 minute. Lower the heat to medium; then stir in the puréed asparagus mixture and the dissolved milk powder. Stir the soup frequently until it is heated through.

Meanwhile, put the reserved asparagus tips into a very small saucepan with a small amount of water and bring to a boil over high heat. Lower the heat and simmer the tips for about 3 to 5 minutes, or until they are just tender; drain. To serve the soup, ladle it into bowls and top each serving with some of the asparagus tips.

Makes about 6 servings.

ASPARAGUS SALAD

Dressing
¼   cup fresh lemon juice
⅓   cup vegetable oil
⅛   teaspoon black pepper, preferably freshly ground
½   teaspoon celery salt
¼   cup finely chopped fresh parsley leaves
¼   cup coarsely chopped sweet red pepper (if unavailable, substitute sweet green pepper)
2   tablespoons grated Parmesan cheese

Vegetables
2   pounds fresh asparagus spears
    Lettuce leaves

Combine all the dressing ingredients in a small bowl and stir to mix well. Set aside for at least 20 minutes to allow the flavors to blend.

Wash the asparagus well. Then gently break off and discard the tough white part at the bottom of each stem. Lay the asparagus spears in a large skillet and cover them with water. Cover the skillet and bring the water to a boil. Lower the heat and gently simmer the asparagus spears just until they are crisp-tender, about 5 to 10 minutes. Do not overcook the spears or they will become mushy and stringy. Drain and cool the asparagus slightly. Arrange the spears on a bed of lettuce leaves. Pour the dressing over the asparagus spears, cover, and refrigerate for at least 30 minutes before serving.

Makes 6 to 7 servings.

SESAME ASPARAGUS WITH LEMON-BUTTER SAUCE

(side dish)

This takes only a few minutes to prepare but looks and tastes quite elegant.

    1   pound fresh asparagus spears, approximately ½ inch thick
    1½   tablespoons sesame seeds
    1½   tablespoons butter or margarine
    1   tablespoon lemon juice, preferably fresh
    ⅛   teaspoon salt (optional)

Wash the asparagus well; then gently break off and discard the tough white part at the bottom of each stem. Lay the asparagus spears in a large skillet and cover them with water. Cover the skillet and bring the water to a boil over high heat. Lower the heat and gently simmer the asparagus spears just until they are tender, about 7 to 10 minutes. Do not overcook the spears or they will become mushy and stringy.

Meanwhile, toast the sesame seeds by stirring them in a small skillet over medium heat, or by heating them on a plate in a microwave oven just until they are golden brown and aromatic. Set aside.

Melt the butter in a small saucepan over medium heat or in a small bowl in the microwave oven. Remove from the heat and stir in the lemon juice and salt (if used).

Drain the cooked asparagus very well; then carefully transfer the spears to a serving platter. Pour the lemon-butter sauce on top of the spears; then sprinkle them with the toasted sesame seeds.

Makes about 4 servings.

ASPARAGUS-NOODLE BAKE

(light main dish or side dish)

Serve this as a luncheon entrée or as a rich side dish with an elegant meal.

3	cups uncooked medium-wide egg noodles
1	tablespoon butter or margarine
1	tablespoon finely chopped onion
1	pound fresh asparagus, trimmed and cut into ½-inch-long pieces (or 1 10-ounce package frozen cut asparagus, thawed)
2¼	cups whole milk, divided
5	ounces sharp Cheddar cheese, grated or shredded (1¼ cups packed), divided
½	teaspoon salt
¼	teaspoon powdered mustard
⅛	teaspoon black pepper, preferably freshly ground
3	large eggs

Cook the noodles in a large pot of boiling water for 6 to 8 minutes, or until cooked through but still slightly firm. Turn them out into a colander and let them drain.

Melt the butter in a medium-sized saucepan over medium-high heat. Add the onion and cook, stirring frequently, for 3 to 4 minutes, or until it is tender. Add the asparagus and 3 tablespoons of water and bring the mixture to a boil. Lower the heat to medium and cook the asparagus pieces, stirring occasionally, for 7 to 9 minutes, or until they are crisp-tender. (If frozen asparagus is used, add only 1 tablespoon of water to the saucepan and reduce the cooking time to about 2 minutes.) Add 1¼ cups of the milk, 1 cup of the cheese, salt, mustard, and pepper to the saucepan. Heat the mixture, stirring frequently, until the milk is hot and the cheese melts. Remove the pan from the heat. With a fork, beat together the remaining 1 cup of milk and eggs until the mixture is smooth. Stir the milk-egg mixture into the asparagus mixture until well combined.

To assemble the casserole, spread the noodles in the bottom of a greased 9- or 10-inch-square flat baking dish. Pour the asparagus mixture evenly over the noodles. Sprinkle the top of the casserole with the remaining ¼ cup of the cheese. Bake the casserole, uncovered, in a preheated 350-degree oven for 45 to 55 minutes, or until the top is nicely browned and the center seems "set" when tapped. Allow the casserole to cool on a rack for about 5 minutes before serving. If using the casserole as a main dish, cut it into squares; if using as a side dish, either cut it into squares or serve by the spoonful.

Makes 4 to 6 main-dish servings or 6 to 8 side-dish servings.

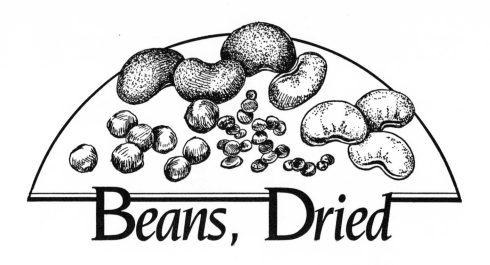

Beans, Dried

One of man's most ancient foods, beans are a member of the legume family, which also includes peas, lentils, and peanuts. All these plants have one basic feature in common: Their roots store nitrogen from the air and turn it into protein.

Although there are more than 10,000 members of the legume family, relatively few are of culinary interest. Some, such as clovers, vetches, and alfalfa, are important as animal fodder.

The ancient Chinese, Egyptians, Greeks, and Romans all cultivated various species of legumes. And several kinds are mentioned in the Bible. Esau exchanged his birthright for lentil potage. When Daniel was brought to the Babylonian court of Nebuchadnezzar, he refused to eat meat, preferring to subsist on a diet of dried peas and water.

Kidney beans are native to the Americas and were an important food for many Indian tribes. Often, they were eaten along with corn, and the two together provided a high-quality source of protein.

While the Indians could not make a scientific study of nutrition, they did apparently stumble upon an important dietary principle. No vegetable source by itself provides "complete" protein. Although beans are high in protein, they do not supply all the essential amino acids in the right quantities necessary to sustain life. However, corn is high in the essential amino acids in which beans are deficient.

17

Eaten in combination, the two foods provide protein as good as that from animal sources. The Indians seem to have recognized this. In fact, friendly tribes taught early settlers how to plant beans and corn together so that the bean vines used the corn stalks for support.

Lima beans and peanuts are also New World natives. The latter were carried by Portuguese explorers to Africa, where they spread so rapidly that they were subsequently mistaken for an indigenous plant.

Legumes were so important in the diet of early man that they were written into the folklore of many peoples. Jack and his famous Bean Stalk and the Princess and the Pea are two well-known examples.

Interestingly, the ancient Greeks and Romans used beans for voting. A white bean signified approval, a black bean disapproval. This is probably the origin of the English custom of handing out a blackball—or "blackballing" someone seeking membership in a club or organization.

Dried legume varieties generally available in the U.S. include the following:

Black or turtle beans are small, dark, and oval with a relatively strong flavor. They are used chiefly in Latin American cooking. They cannot be used interchangeably with other beans because they darken the liquid in which they are cooked.

Black-eyed peas are small and kidney-shaped with a black dot in the center of the inwardly curved side; their unusual flavor is popular in Southern cookery. (They are also sold fresh and frozen.)

Chick-peas are also called garbanzo beans. Medium-sized, round, and rough, they are different in general appearance from other beans and have a slightly crunchy texture and nut-like flavor. They are used in Mediterranean and Mexican cooking. These beans need long cooking.

Cranberry beans are small and red or pink with darker specks. They are similar to pinto beans.

Great Northern beans are large, white, oblong beans with a distinctive flavor. They are good in salads as well as cooked dishes.

Kidney beans are large, red, and (as the name implies) kidney-shaped. Used extensively in Mexican cookery, they are also good in salads.

Lentils are small round flat legumes. Brown lentils (which are greenish-brown to brown in color) are the only ones widely available in the U.S., although there are many other varieties that are used extensively in Indian cuisine. These can sometimes be found at specialty food shops. Lentils cook more quickly than other legumes.

Lima beans, when dried, are light buff or pale green in color and relatively flat. Baby limas are much smaller than the regular ones. Large limas are sometimes called "butter beans." (These beans are also available fresh and frozen.)

Marrow beans are large, round, and white. In fact, they are the largest of the white beans. In the U.S., they are grown primarily in the East.

Navy beans are similar in shape to Great Northern beans, but smaller. They are often used in soups and New England-style "baked bean" dishes.

Pea beans are the smallest of the white beans. Like navy beans, they are used in New England cooking.

Pinto and pink beans are both kidney-shaped beans used in Mexican cookery. They are similar in flavor. Pintos are smooth and off-white with tiny brown specks. Pink beans are uniform in color. Despite the name, they are actually brownish red.

Red beans are sweet and dark red. They resemble kidney beans and are used primarily in Southern cooking.

Soybeans are tiny, round, and firm. An important world agricultural crop, these beans are particularly high in protein.

Split peas are either green or yellow dried peas that have been split in half. They are used primarily in soups.

Availability: Dried beans are available all year round. Because they require a relatively long preparation time, it is sometimes convenient to purchase them canned and already cooked. However, you are paying for the convenience. When cooked, 1 pound of dried beans swells to yield 5 to 7 cups. A pound of canned beans, which costs far more than a pound of dried beans, yields a little more than 1½ cups.

Choosing the Best: Look for clean, well-shaped dried legumes that are not discolored or broken. Avoid those that show any signs of mold.

Nutritional Value: Beans are high in protein. When combined with corn or rice, they are an excellent protein source. They also have significant amounts of iron, calcium, and potassium, as well as thiamine and riboflavin. They are also high in fiber.

Storage: Dried beans can be kept in the pantry for up to a year. After opening the packages, store the beans in closed containers. Cooked beans can be frozen for up to a month.

Basic Preparation and Cooking: Beans should be sorted and well washed before cooking. Remove any broken or discolored ones and foreign objects, such as pebbles. It isn't absolutely necessary to soak beans before cooking. They can be added to soups, for example, directly after being sorted and washed, although they will tend to cook more slowly and less evenly than beans that have been soaked.

To soak beans, cover them with water—generally about 6 to 8 cups of water for a pound of beans. Bring to a boil over high heat and boil for 2 minutes. Remove the pot from the heat and let it stand, covered, for 1 hour. Then drain and rinse the beans before cooking. The beans can also be covered with the 6 to 8 cups of water and soaked at room temperature overnight.

The cooking time of beans varies considerably, depending on variety, although the general range is between 1½ and 2½ hours. Always cook thoroughly as under-done beans are difficult to digest. When done, the beans should be tender but should hold their shape.

Acid foods, such as tomatoes, wine, lemon juice, and vinegar, will increase bean cooking time and, in high concentrations, will prevent the beans from ever becoming tender.

Add flavor by cooking dry beans with chopped onion, garlic, green pepper, and celery, and with herbs, such as bay leaf, thyme, and marjoram. To keep down foam during bean cooking, add a tablespoon of butter or vegetable oil to the cooking water.

Serving Suggestions: Beans are good not only in soups and casseroles, but also marinated or simply chilled and used in salads.

SPICY BLACK-EYED PEA AND BEAN SOUP

½   cup dry navy beans, sorted and washed
½   cup dry black-eyed peas, sorted and washed
3   cups water
1   tablespoon butter or margarine
2   garlic cloves, minced
1   medium-sized onion, finely chopped
5   cups water
2   cups vegetable bouillon (reconstituted from cubes or granules) or vegetable stock (page 86)
¼   cup pearl barley
¼   cup dry lentils, sorted and washed
1   15-ounce can tomato sauce
1   medium-sized carrot, thinly sliced
1   celery stalk, including leaves, coarsely chopped
1   tablespoon packed light brown sugar
2   bay leaves
½   teaspoon chili powder
½   teaspoon powdered mustard
¼   teaspoon dried thyme leaves
⅛   teaspoon ground cloves
¼   teaspoon salt
⅛   teaspoon black pepper, preferably freshly ground

In a large heavy pot, combine the navy beans and black-eyed peas with the 3 cups of water. Bring to a boil, uncovered, over high heat. Lower the heat slightly and boil for 2 minutes. Remove from the heat. Immediately cover and let the beans stand for 1 hour. Drain the beans in a colander.

In the pot in which the beans were soaked, melt the butter over medium-high heat. Add the garlic and onion and cook, stirring constantly, until the onion is soft. Add all the remaining ingredients, including the drained beans. Bring the soup to a boil. Then cover, lower the heat, and simmer, stirring occasionally, for about 2 hours, or until the beans are tender.

Makes 6 to 7 servings.

HEARTY MEATLESS MINESTRONE

This easy recipe makes enough to satisfy a large group; and leftovers are great.

2	tablespoons olive or vegetable oil
2	medium-sized onions, finely chopped
1 to 2	garlic cloves, minced
5	cups water
2	cups vegetable bouillon (reconstituted from cubes or granules) or vegetable stock (page 86)
1	cup dry red wine (or substitute 1 additional cup bouillon)
2	14-ounce cans, or 1 28-ounce can, Italian-style pear-shaped or plum tomatoes, including juice, coarsely chopped
1	8-ounce can tomato sauce
2	medium-sized carrots, diced
2	medium-sized potatoes, peeled and diced (about 2 cups)
1	celery stalk, including leaves, thinly sliced
2 to 3	cups shredded green cabbage
1	9- to 10-ounce package frozen Italian green beans or regular-cut green beans, thawed
1	small yellow squash or zucchini, diced
1	15- to 16-ounce can white beans (any type), well drained
1	15- to 16-ounce can red kidney beans, well drained
1	teaspoon dried basil leaves
1	teaspoon dried marjoram leaves
¼	teaspoon dried thyme leaves
⅛	teaspoon salt
⅛	teaspoon black pepper, preferably freshly ground
1	cup uncooked regular, soy-enriched or whole wheat small elbow macaroni, small shells, or similar pasta
	About ⅓ to ½ cup grated Parmesan cheese (optional)

Heat the oil in a 6-quart or larger soup pot over medium-high heat; then cook the onions and garlic until they are tender but not browned. Stir in the water, bouillon, and wine; then bring the mixture to a gentle boil. Add the tomatoes and their juice, the tomato sauce, carrots, potatoes, and celery. Lower the heat and simmer, covered, for about 20 minutes.

Then add the cabbage, green beans, squash, canned beans, herbs, salt, pepper, and macaroni. Simmer, stirring occasionally, for about 10 minutes longer, or until all the vegetables and the pasta are tender.

If desired, sprinkle each individual serving with about 2 teaspoons grated Parmesan cheese.

Makes 8 to 10 servings.

HERBED WHITE BEAN SALAD

2   15- to 16-ounce cans white beans, drained (or 3 cups cooked and drained navy, pea, or Great Northern beans)
½   cup thinly sliced scallions, including green tops
½   cup finely chopped fresh parsley leaves
3   celery stalks, diced
1   medium-sized sweet red or green pepper, diced

Dressing
¼   cup olive oil
3   tablespoons white wine vinegar
1   teaspoon dried mint leaves (or 1 tablespoon chopped fresh mint leaves)
1   teaspoon dried basil leaves (or 1 tablespoon chopped fresh basil leaves)
¼   teaspoon salt
⅛   teaspoon black pepper, preferably freshly ground

Put the beans, scallions, parsley, celery, and sweet pepper in a bowl and mix gently to combine.

Put all the dressing ingredients in a small jar, cover, and shake well. Pour the dressing over the bean mixture and toss everything lightly until mixed. Cover and refrigerate the salad for at least 1 hour before serving to give the flavors a chance to mingle. Stir the salad occasionally during chilling and just before serving.

Makes about 6 servings.

CHEESY CONFETTI SOYBEANS

(side dish)

Nutritious enough to serve as a light main dish, this takes only a few minutes to prepare. It's also a good way to introduce your family to soybeans—which contain more usable protein than any other bean.

2   tablespoons butter or margarine
2   medium-sized onions, finely chopped
2   garlic cloves, minced
1   medium-sized sweet green pepper, finely chopped
½   medium-sized sweet red pepper, finely chopped (or ½ cup drained and chopped canned pimiento)
2   celery stalks, thinly sliced
2   15- to 16-ounce cans soybeans, drained (or about 3 to 3½ cups cooked and drained soybeans)
⅛   teaspoon black pepper, preferably freshly ground
4   ounces Swiss, mild Longhorn, or Muenster cheese, grated (1 cup packed)

In a medium-sized saucepan over medium-high heat, melt the butter. Cook the onions and garlic until they are tender but not browned. Add the green pepper, red pepper, and celery and cook for 2 to 3 minutes longer, or until they are just tender. Then add the soybeans and black pepper and stir until the soybeans are completely heated through. Finally, add the cheese and stir over low heat until the cheese has melted.

Makes about 6 servings.

BURRITO CASSEROLE

(main dish)

Filling
¾   pound lean ground beef
1   medium-sized onion, finely chopped
1   garlic clove, minced
1   medium-sized carrot, grated or ground
1   15-ounce can tomato sauce
1   4-ounce can (net weight) chopped green chilies, including juice
1   15- to 16-ounce can red kidney beans, well drained and mashed
1   tablespoon chili powder (or to taste)
2 to 3   drops Tabasco sauce (optional)

Burritos and Topping
10   flour tortillas
1   8-ounce can tomato sauce
½   teaspoon chili powder
3   ounces mild Cheddar or Monterey Jack cheese, grated (¾ cup packed)

In a medium-sized skillet over medium-high heat, cook the ground beef, onion, and garlic, breaking up the meat with a spoon, until the meat is brown and the onion is tender. Drain off and discard the excess fat. Add all the remaining filling ingredients to the skillet and stir to mix well. Lower the heat and simmer, uncovered, for about 10 minutes.

Grease a 7- by 11-inch or slightly larger glass baking dish or spray with nonstick vegetable coating. Lay the tortillas one at a time on a plate. Spoon about 3 tablespoons of the filling into each tortilla. Carefully roll up each tortilla and lay side by side, seam side down, in the baking dish.

For the topping, mix together the tomato sauce and the chili powder. Pour this over the rolled tortillas, spreading evenly with a spoon. Sprinkle the top evenly with the grated cheese.

Cover the baking dish with aluminum foil. The casserole can be made to this point and refrigerated. Bake in a preheated 350-degree oven for 30 minutes, or until well heated (or 40 minutes if the casserole has been refrigerated).

Makes about 5 servings.

BEEF BAKE WITH BEANS AND BARLEY

(main dish)

2	tablespoons butter or margarine
1	large onion, finely chopped
2	celery stalks, thinly sliced
2	garlic cloves, minced
1	pound lean stew beef, cut into ½-inch cubes
¼	teaspoon black pepper, preferably freshly ground
1	teaspoon dried thyme leaves
1	teaspoon dried basil leaves
1	cup pearl barley
¾	cup dry pinto or navy beans, sorted and washed
¼	cup dry black-eyed peas, sorted and washed
5½	cups water
1	8-ounce can tomato sauce
1	carrot, grated
1	broccoli stem, grated (Reserve the flowerets for another use.)
1	carrot, thinly sliced
½	cup finely chopped fresh parsley leaves
2	beef bouillon cubes
2	bay leaves
1	teaspoon sugar
1	teaspoon powdered mustard
	Scant ½ teaspoon chili powder
	Pinch of cayenne pepper
½	teaspoon salt

In a large skillet, melt the butter over medium heat. Add the onion, celery, and garlic. Cook, stirring occasionally, until the onion is tender. Push the vegetables to the side of the pan. Add the beef and sprinkle with the black pepper, thyme, and basil. Cook, stirring frequently, until the cubes are lightly browned on all sides.

In a 3-quart casserole, combine all the remaining ingredients. Stir in the cooked vegetables and the beef cubes. Cover the casserole and bake in a 350-degree oven for 2 hours and 20 minutes to 2 hours and 30 minutes, or until most of the water has been absorbed and the beans are tender. Stir several times during cooking.

Makes 8 to 10 servings.

Beans, Fresh

Many of the beans that are eaten fresh
actually belong to the same species as beans that are consumed dried. For example,
green beans are members of the large kidney bean family. However, instead of
being grown for their mature seeds, green beans are cultivated for their seed pods
and eaten while these are still tender and immature.

Green beans—or snap beans as they are properly called—represent a major suc-
cess story for horticultural research. Up until the nineteenth century, snap beans
(including green, wax, and pole types) had strings along their pods, just as other
members of the kidney bean family do. In the varieties grown specifically for their
seeds (Great Northern and navy beans, for example), this tough line of fiber was
useful because it acted as a seam that split and released the plump, edible beans
inside. However, in the kinds of beans eaten pod and all, the strings not only served
no useful purpose, but detracted from the vegetables' appeal.

In the late 1800s, American plant scientists became interested in the "stringi-
ness" problem and set to work to build a better green bean. One pioneer researcher
of the period succeeded in producing nine different types, all tasty and all "stringless." Over time, other experts followed up with improvements of their own, and
today *all* snap beans are stringless at the harvesting stage. Popular nomenclature
still hasn't caught up with the horticultural advances, however; most of us still call

these vegetables "string" beans! (Experts refer to them as "snap" beans because they break with a snap.)

Despite the strings of the early, unimproved green beans, they were well known and enjoyed in the New World long before Columbus reached its shores. (Columbus himself commented that the fields of beans he encountered here were quite different from those in Spain.) It is thought that these edible-podded legumes originated in Central or South America and gradually spread northward. By the 1500s, numerous varieties were growing in North America and had been given native Indian names.

Many "new" edible-podded bush and pole beans—or "foreign beans" as one German source called them—began appearing in European gardens in the mid 1500s, strongly suggesting that they were introduced to the continent from the New World. By the 1600s, a large number of types were described.

Today, the snap bean is among our most popular vegetable crops. Although the medium-green-colored round bean often sold in supermarkets may be considered the standard in the United States, a delicious (but confusing) assortment of less familiar types is also grown. These include round or flat yellow-colored wax; long, flat pole; and wide-podded green bush beans. (The "Italian" green beans seen in many frozen food sections are in fact a wide-podded variety of green bush bean.)

Besides the numerous snap-type beans grown for their edible pods and eaten fresh, several kinds cultivated specifically for their seeds are also sometimes eaten fresh. Called "shell" beans, these include lima beans and black-eyed peas, as well as some lesser known types such as fava and horticultural beans. Many of the shell beans—most notably limas and black-eyed peas—are also good dried. (See the chapter on dried beans for details.)

Availability: Fresh, round green snap beans are available year round throughout the U.S. The peak season is May through August. Flat pole and wax beans are sporadically available in summer, most often in farmers' markets and produce stores. Frozen snap beans, including whole, cut, and French-cut green and Italian green, are widely sold in supermarkets. Canned green and wax beans are likewise standard supermarket items.

Choosing the Best: Fresh snap beans should be firm and free from scars or discoloration. They should be crisp and brittle enough to "snap" when broken in half; pass up beans that are limp or withered. Also, avoid overly mature beans with very large seeds and swollen pods, as these are usually tough.

Nutritional Value: Snap beans are a good source of vitamin C and also contain some vitamin A and potassium.

Storage: Use snap beans promptly, as they will quickly become tough, discolored, and tasteless. If absolutely necessary, they can be kept for 2 or 3 days in the refrigerator in a closed plastic bag. Do not wash the beans until just before using.

Basic Preparation and Cooking: Wash and drain the beans in a colander. Trim off the stem ends with a sharp knife. If desired, trim off the tapered "tail" ends; however, this is not necessary. The beans may be used whole, broken into lengths, or cut diagonally, depending on the recipe. They may be eaten completely raw, although they are more commonly blanched first to bring out their bright green color. To blanch a pound of whole 4- to 5-inch-long beans, bring 3 to 4 quarts of water to a full boil over very high heat. Gradually add the beans and let the water return to a rolling boil. Boil, uncovered, for 4 to 6 minutes, or until the beans are bright green and cooked through but still quite crisp. Turn out into a colander and drain. To set the color of the beans, immediately rinse them well under cold water and drain again. They are ready to be chilled or reheated with seasonings, as desired.

To cook 1- to 1½-inch cut snap beans, combine in a saucepan with just enough water to cover. Bring the water to a boil and simmer for 2 minutes. Then lower the heat, cover the pot, and continue cooking for 12 to 18 minutes longer, or until they are crisp-tender. (Whole beans can be cooked in the same manner, but will take slightly longer.)

Snap beans may also be steamed in a steamer basket over boiling water. Whole beans will take between 16 and 23 minutes, depending on the size; cut beans will take between 15 and 21 minutes.

Simple Serving Suggestions: Blanched, chilled green beans are excellent marinated in viniagrette dressing or added to a salad bowl. Serve cooked beans dressed with butter, salt, and pepper; sautéed mushrooms; toasted slivered almonds or sesame seeds; or a dash or two of soy sauce.

ITALIAN GREEN BEANS AND WATER CHESTNUTS SALAD

The use of frozen Italian beans and canned sliced water chestnuts in this salad makes it very quick and convenient. The water chestnuts add crunch and color contrast to the beans, which are flatter and wider than regular green beans.

Vegetables

2 9-ounce packages frozen Italian green beans
1 8-ounce can sliced water chestnuts
2 tablespoons finely chopped fresh parsley leaves

Dressing

3 tablespoons olive oil
3 tablespoons red wine vinegar
1 teaspoon instant minced onions
½ teaspoon powdered mustard
   Scant ½ teaspoon salt
⅛ teaspoon black pepper, preferably freshly ground
⅛ teaspoon garlic powder
⅛ teaspoon sugar

Cook the green beans following package directions, just until they are crisp-tender. Then drain them in a colander and rinse them briefly with cold water to stop the cooking. Drain the beans very well.

Drain the canned water chestnuts; then rinse them with cold water and drain completely. Put the beans and water chestnuts into a medium-sized bowl with the parsley.

Combine the dressing ingredients in a small jar with a tight-fitting lid; then cover it tightly and shake well. Pour the dressing over the vegetables and toss them to mix.

Cover and refrigerate the salad for several hours or overnight, stirring occasionally, to give the flavors a chance to mingle. Serve the salad chilled or at room temperature.

Makes about 5 servings.

GREEN BEANS WITH PIZAZZ

(side dish)

This easy recipe turns green beans into a colorful and satisfying side dish that will add interest to any meal. Leftovers are good served chilled as a salad.

1	pound fresh green beans, stem ends removed
1½	tablespoons butter or margarine
1	medium-sized onion, finely chopped
1	garlic clove, minced
1	medium-sized sweet green or red pepper, diced
1	16-ounce can tomatoes, including juice, chopped
½	teaspoon dried basil leaves
¼	teaspoon salt
¼	teaspoon black pepper, preferably freshly ground

Cut or break each green bean in half. Set aside. In a large skillet over medium-high heat, melt the butter; then cook the onion and garlic, stirring often, until they are tender but not browned. Add the green pepper and green beans, and cook, stirring, about 2 minutes longer. Stir in the tomatoes and their juice, basil, salt, and pepper.

Bring the mixture to a boil; then lower the heat and cover the skillet. Simmer for 15 minutes; then uncover the skillet and continue cooking the green beans, stirring often, about 5 minutes longer, or until they are tender and most of the liquid has evaporated.

Makes about 6 servings.

GREEN BEANS WITH SESAME SEEDS

(side dish)

Very easy, yet delicious!

2 tablespoons sesame seeds
1 small onion, finely chopped
1 garlic clove, minced
1 tablespoon peanut or vegetable oil
1 pound fresh green beans, trimmed and broken in half crosswise
⅓ cup water
½ teaspoon lemon juice
⅛ teaspoon salt
⅛ teaspoon black pepper, preferably freshly ground

Spread the sesame seeds in a large skillet over high heat. Heat, stirring, for 4 to 5 minutes, or until the seeds are lightly browned. Remove the seeds from the skillet and set them aside.

Combine the onion, garlic, and oil in the same skillet. Cook the vegetables, stirring, for 1½ minutes, or until they are lightly browned. Add the beans to the skillet and cook, stirring, for 1 minute longer. Add all the remaining ingredients, *except* the reserved sesame seeds, and allow the mixture to return to a simmer. Cover the skillet and simmer the beans for 10 to 12 minutes, or until they are almost tender. Uncover the skillet and continue cooking until the beans are tender and all the excess liquid has evaporated from the pan. Sprinkle the sesame seeds over the beans and serve.

Makes 4 to 5 servings.

VEGETABLES IN CREAMY TOMATO SAUCE

(side dish)

Vegetables

2½ cups 1-inch green bean pieces, stem ends removed
1½ cups 1-inch unpeeled "new" red potato cubes
  1 large sweet green pepper, cut into 1-inch squares
  1 medium-sized turnip, peeled and cut into ¾-inch cubes
  1 medium-sized onion, finely chopped
  1 garlic clove, minced

Sauce

  1 8-ounce can tomato sauce
  ¼ cup dry white wine
  ¾ teaspoon dried basil leaves
  ½ teaspoon dried thyme leaves
  ¼ teaspoon dried marjoram leaves
  ⅛ teaspoon powdered mustard
    Pinch of cayenne pepper
  ¼ teaspoon salt
  ⅛ teaspoon black pepper, preferably freshly ground
  ½ cup lowfat milk
  2 tablespoons instant nonfat dry milk powder
  1 teaspoon cornstarch
  1 tablespoon butter or margarine

Combine the vegetables in a large saucepan. Cover with water. Cover the pot and bring to a boil over high heat. Lower the heat and simmer for about 11 to 14 minutes, or until the vegetables are tender. (The string beans will be crisp-tender.) Remove from the heat. Drain the vegetables in a colander. Return the vegetables to the pan.

While the vegetables are cooking, combine the tomato sauce, white wine, basil, thyme, marjoram, powdered mustard, cayenne pepper, salt, and black pepper in a small saucepan. Cover and bring to a boil over medium-high heat. Lower the heat and simmer for 5 minutes.

Meanwhile, combine the milk, dry milk powder, and cornstarch in a small cup. Stir to dissolve the dry milk powder and cornstarch.

Remove the tomato sauce mixture from the heat and stir in the milk mixture and butter. Return the pan to the heat and cook over medium-high heat, stirring, until the sauce thickens and boils. Pour the sauce over the vegetables and stir carefully so as not to break up the potatoes. Return the vegetable pan to the burner and lower the heat to medium. Cook, simmering, for about 5 minutes so that the flavors can blend.

Makes 5 to 6 servings.

LIMA BEANS WITH SWEET RED PEPPER

(side dish)

A quick and tasty way to dress up frozen lima beans.

1½   tablespoons butter or margarine
2   tablespoons finely chopped onion
⅔   cup diced sweet red pepper
¾   cup water (approximately)
⅛   teaspoon salt
⅛   teaspoon black pepper, preferably freshly ground
1   10-ounce package frozen baby lima beans
3   tablespoons grated sharp Cheddar cheese

Melt the butter in a medium-sized saucepan over medium-high heat. Add the onion and sweet red pepper and cook, stirring, for 3 to 4 minutes, or until the onion is limp. Stir in the ¾ cup water and all the remaining ingredients, *except* the cheese, and bring to a boil. Lower the heat, cover, and simmer the mixture for 13 to 16 minutes, or until the beans are just tender and most of the liquid has evaporated from the pan. (If the pan begins to boil dry during cooking, add another tablespoon or two of water.) Transfer the mixture and any remaining liquid to a serving bowl. Sprinkle the top with the grated Cheddar cheese and serve immediately.

Makes 4 to 5 servings.

SAVORY BLACK-EYED PEAS AND RICE

(side dish)

1½ tablespoons peanut or vegetable oil
1   medium-sized onion, coarsely chopped
½   cup diced sweet green pepper
¼   cup diced sweet red pepper (optional)
2   cups chicken broth or bouillon (reconstituted from cubes or granules), divided
1   bay leaf
⅛   teaspoon salt
    Pinch of ground allspice
    Pinch of dried thyme leaves
    Pinch of cayenne pepper
1   10-ounce package frozen black-eyed peas, thawed
½   cup *uncooked* long-grain white rice

Combine the oil, onion, green pepper, and red pepper (if used) in a medium-sized saucepan over medium-high heat. Cook the vegetables, stirring, for 3 to 4 minutes, or until they are limp. Stir in 1 cup of the chicken broth and all the remaining ingredients, *except* the rice. Bring the mixture to a boil. Lower the heat, tightly cover the pan, and simmer gently, without stirring, for 25 minutes. Stir in the remaining 1 cup of chicken broth and the rice and bring the mixture to a simmer once more. Cover the pan and gently simmer the mixture, without stirring, for 15 to 18 minutes longer, or until the rice and peas are just tender and most of the chicken broth has been absorbed. Discard the bay leaf and serve.

Makes 5 to 6 servings.

CHICKEN AND FRESH VEGETABLE SKILLET DINNER

(main dish)

2	tablespoons vegetable oil
1	large onion, finely chopped
1	garlic clove, minced
2½ to 3	pounds meaty chicken pieces
¼	teaspoon salt
⅛	teaspoon black pepper, preferably freshly ground
½	teaspoon dried thyme leaves
½	teaspoon dried basil leaves
1	medium-sized bay leaf, crumbled
1	cup chicken broth or bouillon (reconstituted from cubes or granules)
2	large tomatoes, diced
4	medium-sized carrots, thinly sliced
1½	cups fresh green beans, cut in half crosswise
1	tablespoon cornstarch
2	tablespoons cold water

To serve

Hot cooked white or brown rice, OR regular or whole wheat noodles

In a Dutch oven or very large deep skillet, heat the oil over medium-high heat; then add the onion and garlic and cook until they are tender but not browned. Add the chicken pieces and brown them lightly on both sides. Then stir in the salt, pepper, thyme, basil, bay leaf, and broth and bring to a boil.

Cover the pot tightly, lower the heat, and simmer the chicken for 30 minutes. Stir in the tomatoes, carrots, and green beans; then simmer the chicken and vegetables, covered, for an additional 15 to 20 minutes, or until they are tender.

In a small cup, mix together the cornstarch and water. Add the mixture to the pot, stirring, and cook until the sauce thickens and boils. Serve over hot cooked rice or noodles.

Makes about 4 servings.

SALAD NIÇOISE

(main dish)

Though there are hundreds of variations on this hearty main-dish salad, most seem to include green beans, potatoes, and tuna. Try to use vegetables that are as fresh as possible. They really make this a great summer meal.

The salad has several parts, but all are easily prepared up to a day ahead, and the complete salad can be assembled up to 2 hours before serving.

6 or 7	small "new" red potatoes, scrubbed (about 1¼ pounds)
⅓	cup olive oil
3	tablespoons white wine vinegar
1	tablespoon lemon juice, preferably fresh
1	garlic clove, minced
½	teaspoon powdered mustard
½	teaspoon dried basil leaves
½	teaspoon dried thyme leaves
¼	teaspoon salt
⅛	teaspoon black pepper, preferably freshly ground
2	tablespoons finely minced scallions, including green tops
3 to 4	tablespoons dry sherry or dry white wine
½	pound fresh green beans, stem ends removed
8 to 12	romaine lettuce leaves or curly red lettuce leaves
3 to 4	medium-sized vine-ripened tomatoes, cut into wedges (or 1 pint cherry tomatoes)
3 or 4	hard-boiled eggs, peeled and cut into wedges
	About ¼ cup ripe (black) olives
2	6½-ounce cans solid white water-packed tuna, drained
2 to 3	tablespoons drained capers (optional)

Put the potatoes in a medium-sized saucepan with about 1 inch of water. Bring to a boil over high heat. Cover tightly, lower the heat, and steam the potatoes for about 20 to 30 minutes, or until they are tender.

Meanwhile, prepare the dressing. Put the oil, vinegar, lemon juice, garlic, mustard, basil, thyme, salt, and pepper in a small jar. Cover it and shake well.

When the potatoes are done, immediately drain them. Cool them slightly; then cut them into coarse chunks and put them into a medium-sized bowl. Stir in the scallions; then gently mix in the sherry until it is completely absorbed. Finally, stir in ¼ cup of the dressing. (Reserve and chill the remainder of the dressing to use with the completed salad.) Chill the potato mixture.

Put the green beans in a saucepan with a small amount of water (or use a steamer), and steam them until they are brightly colored and crisp-tender, about 5 to 8 minutes. Then briefly rinse them under cold water to immediately stop the cooking. Drain and chill the green beans.

To assemble the salad: Arrange the lettuce on a large serving platter or tray. Mound the prepared potato mixture in the center. Arrange the chilled green beans in bunches that radiate from the center. Put the tomato and hard-boiled egg wedges in the spaces between the beans. Arrange the olives around the edge of the platter. Put chunks of tuna in any remaining spaces and scatter the capers (if used) over the top. Shortly before serving, sprinkle the remaining dressing over the salad.

Makes 4 to 6 servings.

Beets

Beets, which are probably native to Europe and North Africa, are mentioned in the writings of both ancient Greek and Roman gourmets. The early Romans ate only the leaves. The roots were reserved for a variety of medicinal purposes. Beet root was thought to relieve headache and toothache. And pieces were sometimes also inserted into the nose like snuff to bring about sneezing. (For this purpose, strong smelling white beet was preferred!)

Although recipes for beet root are found in Roman literature from the early Christian era, there is little further mention of the vegetable until the Renaissance. And beets were not widely cultivated in Europe until the nineteenth century.

Beets can be red, yellow, or almost white in color. The red types are much more common in America, the lighter varieties in Europe. While table beets are cultivated for their roots, the leaves, or greens, are also edible and taste a bit like spinach. Indeed, both species are members of the Goosefoot family.

Sugar beets, a large tapered variety with white flesh and almost white skin, were initially grown chiefly as animal fodder. At the beginning of the nineteenth century, a German chemist found that sugar could be extracted from them. But there was little commercial interest in the discovery until the Napoleonic era when the English blockaded French ports and cut off the supply of cane sugar. Napoleon ordered 70,000 acres to be planted in sugar beets, and a refinery to process them was built in Paris. It was able to supply both the royal palace and Parisian consumers with the new sugar. Today, sugar beets supply about a third of the sugar consumed in Europe.

Swiss and red chard both belong to the same family as the beet, although these are cultivated strictly for their edible foliage.

Availability: Beets are most readily available from June through October, although they can be purchased year round in some parts of the country.

Choosing the Best: Look for beets that are well rounded, firm, and smooth skinned. Small to medium-sized beets will be more tender than larger ones. If the beets are sold untrimmed, their leaves should be crisp and deep green with red veins.

Nutritional Value: Beet roots are high in potassium. The greens are a good source of iron, calcium, and vitamin A. In addition, beets are low in calories and high in fiber. There are about 32 calories in 3½ ounces of cooked beets. And 3½ ounces of cooked beet greens have only 18 calories.

Storage: Beets can be kept for up to a week in the refrigerator. They should be placed in the crisper compartment or in a plastic bag. They remain freshest if stored loosely packed and unwashed.

Preparation and Basic Cooking: Beets should be scrubbed thoroughly but gently before cooking. To preserve their color, do not peel or break the skin.

Beets are most commonly boiled. To cook them whole, cut off all but 1 to 2 inches of the stem; leave the root intact. Cover the beets with water and bring to a boil. Cover the pot, lower the heat, and simmer for about 20 to 35 minutes, or until the beets are tender, depending on size. Cool the beets slightly and trim off the roots and stems with a knife. Using your fingers, slip off the skins.

For quicker cooking, beets may be sliced or shredded. However, some of the color will be lost into the cooking water. Slices should be simmered for about 10 to 20 minutes; shredded beets for about 5 to 10 minutes.

The coarsely chopped stems and leaves can be simmered along with the sliced beet roots for 15 to 20 minutes. In addition, the greens can be cooked separately in water for 15 to 20 minutes, or until they are tender. (Coarse leaves should be shredded first.)

Simple Serving Suggestions: Cooked beets can be seasoned with a little salt, pepper, and butter. Grated orange or lemon rind, lemon juice, vinegar, dill, savory, and thyme also complement their flavor. Uncooked beet greens are excellent in salads.

BORSCHT

Beet lovers will appreciate this unusual, brightly colored vegetarian soup.

2   tablespoons butter or margarine
2   cups finely chopped onion
1   garlic clove, minced
2   celery stalks, thinly sliced
4   cups water
2   cups vegetable boullion (reconstituted from cubes or granules) or vege-
    table stock (page 86)
2   medium-sized carrots, grated or shredded
1   medium-sized turnip, peeled and grated or shredded
2½ cups peeled and shredded beets
2   cups grated or shredded green cabbage
1½ cups peeled and grated or shredded potato
2   tablespoons apple cider vinegar
3   bay leaves
½   teaspoon dried thyme leaves
¾   teaspoon salt
½   teaspoon black pepper, preferably freshly ground

In a large heavy pot, melt the butter over medium-high heat. Add the onion, garlic, and celery. Cook, stirring, until the onion is tender. Add the remaining ingredients. Bring to a boil. Cover, lower the heat, and simmer for about 40 to 45 minutes, or until the vegetables are tender and the soup has thickened slightly. Remove about half the vegetables and liquid and purée in a food processor or blender, in batches if necessary. Return the puréed vegetables to the pot and stir well to blend.

Makes 7 to 8 servings.

BEETS AND GREENS

(side dish)

1   bunch (5 to 6) 1½- to 2-inch-diameter beets with greens
    Generous ⅛ teaspoon salt
⅛   teaspoon black pepper, preferably freshly ground
½   tablespoon butter or margarine

Carefully trim off the greens from the beets, leaving about 1 inch of stem on the top of each. Leave the beet roots intact. Reserve the greens in a colander. Using a vegetable brush, gently scrub the beets under cool running water; scrub thoroughly but be careful not to puncture the skins or the beets will "bleed" excessively.

Put the whole beets into a medium-sized saucepan and add enough water to come ½ inch up the pan sides. Add the salt and pepper to the pan. Tightly cover the pan and bring the water to a boil over medium-high heat. Lower the heat and simmer, tightly covered, for 10 minutes.

Meanwhile, prepare the greens by washing them well in several changes of cool water. Cut the beet stems into 2-inch lengths. Discard any dry, very large, or browned leaves; tear the remaining leaves into bite-sized pieces. Rinse and drain the stem pieces and torn leaves in a colander.

Stir the prepared stem pieces and leaves into the pan; at first the greens will fill the pan but they will gradually wilt and fit easily. Replace the pan lid and simmer the beets and greens for 10 to 15 minutes, or until the stem pieces are almost tender and the beets "give" slightly when pressed with a fingertip. Remove the pan from the heat. Remove the whole beets from the pan and let them cool in a colander for several minutes. Trim off the tops and roots using a sharp knife. If desired, slip off the beet skins with your fingers or remove them with a vegetable peeler. Cut the beets into quarters or ¼-inch-thick slices and return to the pan. Add the butter to the pan and return to the heat. Continue simmering, uncovered, until the butter melts and the stem pieces and cut beets are just barely tender. Transfer the beets, greens, and any pot liquid to a serving bowl or to individual dishes.

Makes 4 to 6 servings.

SWEET AND SOUR SAUCED BEETS

(side dish)

2   large bunches beets (about 2 pounds without greens)
2   tablespoons butter or margarine
1   tablespoon cornstarch
2   tablespoons apple cider vinegar
2   tablespoons packed light or dark brown sugar
½   teaspoon ground cinnamon
¼   teaspoon ground cloves
Pinch of salt

Cut off the beet stems about 1 inch above the beets and leave the roots intact. Scrub the beets well, being careful not to break the skins. Put the beets into a medium-sized saucepan and cover them with water. Bring to a boil over high heat; then cover the pan, lower the heat, and simmer the beets for about 20 to 30 minutes, or until they are very tender. (They should "give" slightly, when pressed with a fingertip.)

Let the beets cool in the cooking water until they can be handled. Drain the beets, reserving ½ cup of the cooking liquid. Carefully cut off the roots and stems and slide the skin off the beets. Cut the beets into ¼-inch-thick slices and set them aside while preparing the sauce.

For the sauce, put the butter in the saucepan and melt it over medium-high heat. Meanwhile, dissolve the cornstarch in the reserved ½ cup beet liquid and add the vinegar, brown sugar, cinnamon, cloves, and salt. Stir this mixture into the melted butter and cook, stirring, until the sauce thickens and comes to a boil. Add the cooked beet slices and cook, stirring, until the beets are heated through.

Makes about 6 servings.

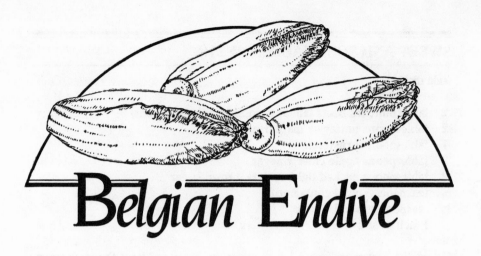

Belgian Endive

If the label "gourmet" is used to signify those foodstuffs that are expensive, time consuming to produce, and in limited supply, then Belgian endive is "gourmet" indeed. Virtually the entire world supply of these exotic, sprout-like leaf heads is grown in one small country—Belgium (surprise!). And it takes 5 to 6 months to yield a single crop.

In fact, the lengthy and rather complicated cultivation process—which the Belgians also developed—helps explain just why this unusual member of the chicory family is so highly prized. First, seeds from the witloof (Flemish for white leaf) variety of chicory are sown in the spring. The plants are allowed to grow and produce leaves all summer. Then, in the fall, the leaves are cut off and the roots dug up and replanted in the dark in a special sandy medium. After a few weeks, the roots begin to put out plump, creamy-white shoots. When the tips of these tight sheaths of leaves reach the surface, the slender heads are "picked." A 10-foot long row of roots may yield only 30 or 40 little endive heads.

The first Belgian endives were grown by accident in the last century when some chicory roots that had been left in a dark spot began to sprout. A Brussels horticulturalist then set to work developing the vegetable in earnest. Later, when a shipment of cultivated endives sent to Paris became a hit, the Belgian endive industry was launched.

Like its close relative, the chicory root used in coffee, Belgian endive has a distinctive, pleasantly bitter taste. The fresh leaves are smooth and wax-like and add an appealing pungency and crispness to salads. Cooked whole, Belgian endives have a tenderness and succulence all their own.

Two other members of the chicory family are curly leaf endive and escarole. Both have a pungent flavor reminiscent of their more aristocratic cousin. In the U.S., they are usually eaten raw as salad greens. However, in Europe and the Near East they may be served as cooked vegetables.

Availability: Almost without exception, the Belgian endives available in the United States are imported from Belgium. Since witloof is primarily a winter vegetable, the supply of endives is greatest from September through April.

Choosing the Best: Select heads that are firm, smooth, compact, and lustrous looking. Except for the tips, which are pale yellowish-green, the shoots should be white. Avoid any that are soft or that have dark spots or brown-tinged leaves or bases.

Nutritional Value: Although curly leaf endive, escarole, and chicories grown for their mature green leaves are good sources of vitamin A, Belgian endives are still undeveloped plant shoots and have little nutritional value. They do provide fiber, however, and, being over 95 percent water, are low in calories. (A whole pound of Belgian endives has only about 60 calories.)

Storage: Belgian endives keep well stored in plastic bags in the crisper. However, for best flavor, use within several days. Do not wash until just before using.

Preparation and Basic Cooking: Belgian endives are most often eaten raw, but they are also a special treat cooked. In either case, they require almost no preparation. Simply wipe or lightly rinse off the heads and trim off a thin layer at the base. If you wish to minimize the bitterness, cut up through the base in a circular motion and remove the core. To use endives raw in salads, gently pull off individual leaves one by one, or cut the heads crosswise into rounds or lengthwise into coarse shreds. When the heads are to be cooked, they are usually left whole or split in half lengthwise.

Simple Serving Suggestions: Raw endive rounds or shreds can be included in any salad bowl where a crunchiness and slightly bitter accent are welcome. Due to their firmness and pungency, endive rounds or shreds stand up well to a zesty vinaigrette dressing. Individual leaves make a pretty and elegant salad bed when arranged tips outward in a fan-like or circular pattern. The leaves can also be used as an elegant appetizer by inserting a dab of filling or special dip into the tip of each. Whole or split heads can be braised or baked in a little beef or chicken broth until they are tender. Or steam them in a steamer basket and serve with a Mornay sauce or sprinkled with Parmesan cheese.

STUFFED BELGIAN ENDIVES

(appetizer)

Filling

½ cup part-skim ricotta cheese (if unavailable, substitute regular ricotta)
¼ cup feta cheese
2 tablespoons finely chopped fresh spinach leaves or parsley leaves
½ teaspoon dried dillweed
2 to 3 drops Tabasco sauce

To Serve

2 medium-sized heads Belgian endive
Paprika for garnish

Combine all the filling ingredients in a small bowl and mix well with a fork, mashing the feta to incorporate it into the ricotta. Cover the bowl and chill the filling for at least 1 hour to allow the flavors to blend.

Trim the bottom of each endive. Remove the leaves one at a time and fill the tip of each with a generous teaspoonful of the filling mixture. (Save the small inner leaves for another use.) Garnish with paprika. Arrange the endive leaves attractively on a serving platter.

Makes ⅔ cup filling or enough for about 2 medium-sized endives.

BELGIAN ENDIVE WITH CREAMY HERB DRESSING

(composed salad)

Once you have acquired a taste for the slightly bitter flavor and crisp texture of elegant Belgian endive, you will want to use its white leaves in salads and as an attractive garnish. Though the tiny heads may seem somewhat expensive on a per pound basis, they are really rather lightweight. And a little goes a long way when each head is separated into leaves.

The slightly rounded shape of each leaf makes it a perfect receptacle for cradling dressing. The one below is thick and rich tasting, but it is low in fat and calories and high in protein and calcium.

Dressing
⅔	cup plain lowfat or regular yogurt
⅓	cup commercial sour cream
2	tablespoons milk (or as needed)
1 to 1½	tablespoons apple cider vinegar (to taste)
1 to 1½	teaspoons sugar (to taste)
¼	teaspoon onion powder
¼	teaspoon garlic powder
⅛	teaspoon celery salt
½ to 1	teaspoon mixed ground dried herbs, such as marjoram, basil, thyme, tarragon, or mint (or your choice, to taste)*

Salad
4	medium-sized, firm, compact heads Belgian endive
1	pint cherry tomatoes, each cut in half

For the dressing, stir or shake all the ingredients together in a small jar. If possible, make the dressing several hours ahead to give the flavors a chance to blend. (The dressing will keep for about a week in the refrigerator. If it is too thick, thin it with additional milk.)

To serve the salad, cut off and discard the root end from each head of Belgian endive, and gently peel off each of the leaves. Using 1 endive head per serving, lay out the leaves on 4 salad plates so that the leaves radiate from the center like the petals of a flower. Smaller leaves may be placed in between larger ones. Place a quarter of the cherry tomato halves in the center of each plate.

Spoon about a teaspoon of the dressing onto the base of each leaf. Pass the remaining dressing at the table.

Makes 4 servings.

*A good, pre-mixed herb combination that contains no added salt is Bouquet Garni, made by Spice Islands.

BAKED BELGIAN ENDIVES

(side dish)

- **4** plump Belgian endives
- **1** tablespoon butter or margarine
- **1½** tablespoons chopped fresh chives (or **2½** teaspoons dried chopped chives)
- **1** tablespoon enriched all-purpose or unbleached white flour
- **⅓** cup beef broth or bouillon (reconstituted from cubes or granules)
- **¼** cup dry white wine
- **3** tablespoons light cream or half-and-half
- **⅛** teaspoon paprika
- **Pinch of ground nutmeg**

With a sharp knife, carefully core the Belgian endives by removing a ½-inch-deep cone of flesh from the bottom of each. Split the endives in half lengthwise. Arrange them, cut side up and tips touching, in 2 rows in a 9- or 10-inch-square (or similar) flat baking dish.

Melt the butter in a small heavy saucepan over medium-high heat. Add the chives and cook, stirring, for 1½ minutes. Stir the flour into the butter-chive mixture until the flour is completely incorporated. Cook, stirring, for 1 minute longer. Slowly add the broth to the saucepan, stirring vigorously until the mixture is smooth and well blended. Stir in all remaining ingredients and allow the mixture to return to a simmer. Cook, stirring, for 2 minutes longer, or until the sauce is smooth and slightly thickened. Pour the sauce over the Belgian endive halves, being sure to moisten each. Tightly cover the baking dish. Bake the endive halves in a preheated 375-degree oven for 17 to 22 minutes, or until they are just tender.

Makes about 4 servings.

Broccoli

Broccoli has been cultivated for at least 2,000 years, and has been widely enjoyed in Italy for centuries. However, until fairly recently, Americans considered it only a minor Italian specialty vegetable. In fact, broccoli had the nickname "Italian asparagus."

In the 1920s, some enterprising California growers of Italian extraction shipped several crates of ice-packed broccoli to Boston, where it was enthusiastically accepted. Soon after that, the vegetable became commercially important throughout the country, and it has remained so ever since.

The name "broccoli" comes from the Italian plural diminutive of *brocco*, which means arm, branch, or shoot. The type most commonly grown and sold in the United States is sometimes called "sprouting broccoli," and has tightly closed, dark green, blue-green, or purplish buds on top of long, dark-colored fleshy stems. Both green and purple broccoli turn bright green when cooked.

Another type of broccoli, which is actually quite obscure, is sometimes called "heading broccoli." It has dense white heads that are virtually the same as, and sometimes even considered a type of, cauliflower.

(The term "broccoli" can also refer to the central flower stalk that develops in overgrown cabbage. Though this meaning has become archaic in the United States, it is still common in France. Similarly, in England, the term "white broccoli" may be used to describe winter cabbage.)

A member of the large cruciferous group, broccoli is a close cousin not only to cauliflower and cabbage, but also to all the other vegetables in the cabbage family.

A related green, broccoli rabe (or di rape) is becoming more popular in the U.S. This vegetable looks like very thin, leafy broccoli, but is sharper in flavor.

Availability: Fresh broccoli is available the year round in most places, though the peak occurs during October through April.

Choosing the Best: The heads of the broccoli should be compact and brightly colored, with buds that are tightly closed. Open buds with yellow flowers are a sign of overmaturity. The stems should be firm, not limp or soft, and have dark green leaves that are not too wilted. Although stalk size does not affect quality, avoid broccoli with very thick stems that have holes in the bottom, as they may be woody and tough.

Nutritional Value: Broccoli is probably one of the most all-around healthful (and readily available) vegetables. It is an excellent source of both vitamins A and C, and is also high in iron, calcium, potassium, and other minerals. Broccoli is also a good source of non-nutritive fiber. All this, and it is only about 40 calories per cup!

(As a member of the cruciferous, or cabbage family, broccoli may also play a role in the prevention of certain types of cancer.)

Storage: Store broccoli loosely wrapped in the refrigerator, and use as soon as possible. Although it may keep for several days, it will eventually lose its crisp texture and take on an unappealing odor and flavor.

Preparation and Basic Cooking: Rinse the broccoli; if desired, remove and discard any leaves on the stalk. (The leaves are nutritious and considered by some to be very tasty. They can be cooked along with the rest of the broccoli.) Fresh broccoli is almost always cut up before it is cooked or eaten raw as a crudité. The following method is easy and produces very little waste. It also allows the stem and flowerets to be evenly cooked together.

For each stalk, cut off the flowerets at the point where their narrow branches reach the main stem and set them aside. Trim off and discard the bottom ½ inch of the stem. Peel and discard the woody covering from the stem. Cut the stem crosswise into thin slices or lengthwise into juilienne strips. The broccoli is now ready to be served raw as a crudité or to be cooked.

To cook the flowerets and stem pieces together, put them in a steamer basket set above simmering water, and steam them for about 8 minutes, or just until they are bright green and crisp-tender. Do not overcook broccoli or it will become mushy and very strong tasting.

Broccoli may also be cooked in boiling water if preferred. In this case, the cooking time is only about 5 minutes.

To stir-fry broccoli, heat a few tablespoons of oil in a wok or large skillet over medium-high heat. Add the cut up broccoli, and stir-fry it for about 2 minutes. Then add a few tablespoons of water or soy sauce and cover the pan tightly. Steam the broccoli for about 4 minutes, or until it is crisp-tender.

   Simple Serving Suggestions: Fresh raw broccoli is delicious served with a dip or vinaigrette, or as part of a tossed salad. If desired, it may first be partially cooked for a few minutes (to tenderize it slightly and bring out the bright-green color and good flavor) and then chilled.
   Cooked broccoli stands on its own with just a bit of butter or lemon juice and it is also quite good with hollandaise sauce.

CURRY DIP

The tangy flavor of this dip is particularly good with fresh broccoli flowerets. For variety, serve it with assorted raw vegetables, such as carrot and celery sticks, green pepper rings or strips, thin cucumber and turnip slices, and cauliflower flowerets.

Dip
½ cup mayonnaise
¼ cup plain lowfat yogurt
¼ cup commercial sour cream
½ teaspoon sugar
½ teaspoon curry powder
¼ teaspoon onion powder
¼ teaspoon black pepper, preferably freshly ground
¼ teaspoon powdered mustard
⅛ teaspoon garlic powder

To Serve
   Broccoli flowerets and other raw vegetables, such as carrot and celery sticks, green pepper rings or strips, thin cucumber and turnip slices, and cauliflower flowerets

In a small bowl, combine all the dip ingredients and stir to mix well. Cover and refrigerate for several hours to allow the flavors to blend. Serve with broccoli flowerets and assorted raw vegetables.
   Makes about 1 cup.

BROCCOLI-CHEESE-PASTA SOUP

A perfect main course for a luncheon or light supper, this hearty and nutritious soup takes only about 20 minutes to prepare. The quickest and easiest way to chop the broccoli is with a food processor; however, it can also be done by hand if necessary.

1½	tablespoons butter or margarine
1	medium-sized onion, finely chopped
1	garlic clove, minced
7	cups water
2	vegetable or chicken bouillon cubes or packets
8	ounces regular or whole wheat pasta twists (about 3 cups)
3 to 4	cups finely chopped fresh broccoli, including flowerets and stems
¼	teaspoon salt
1	cup instant nonfat dry milk powder
1	cup skim, lowfat, or whole milk
12	ounces sharp Cheddar cheese, shredded (3 cups packed)
5	tablespoons enriched all-purpose or unbleached white flour

In a soup pot or Dutch oven, melt the butter over medium-high heat; then cook the onion and garlic until they are tender but not browned. Add the water and bouillon cubes and bring to a boil, stirring occasionally to dissolve the bouillon. Add the pasta, broccoli, and salt gradually, so that the broth does not stop boiling. Cook, uncovered, stirring occasionally, until the pasta and broccoli are almost tender, about 7 minutes.

Dissolve the milk powder in the milk; then stir the mixture into the soup. Toss the cheese with the flour; then add it very gradually to the soup, stirring constantly until the cheese is completely melted and the soup simmers at least 1 minute.

Makes about 8 generous servings.

CURRIED BROCCOLI SALAD

Dressing
- ⅓ cup mayonnaise
- ¼ cup commercial sour cream
- ¼ cup plain lowfat yogurt
- 1 teaspoon instant minced onions
- 2 teaspoons curry powder (or to taste)
- 2 teaspoons sugar
- 1 teaspoon Dijon-style mustard
- ⅛ teaspoon salt

Salad
- 7 to 8 cups broccoli flowerets and peeled, sliced stem pieces
- 1 8-ounce can sliced water chestnuts, well drained
- ¼ cup sliced almonds
- 1 celery stalk, thinly sliced

In a small bowl, mix together all the dressing ingredients and set them aside.

Meanwhile, put the broccoli into a medium-sized saucepan. Add about 1 inch of water. Cover and bring to a boil. Lower the heat and cook the broccoli for about 5 minutes, or until it is brightly colored and crisp-tender. Drain well in a colander.

In a large bowl, combine all the salad ingredients, including the steamed broccoli. Add the dressing and toss to mix well. Cover and refrigerate for at least 1 hour to allow the flavors to blend.

Makes about 6 servings.

SWEET AND SOUR BROCCOLI

(side dish)

Sauce
2	tablespoons sugar
2	tablespoons water
1½	tablespoons red wine vinegar
1	tablespoon soy sauce
1¼	teaspoons cornstarch
	Pinch of ground allspice
	Pinch of ground ginger
	Pinch of cayenne pepper

Vegetables
2	tablespoons peanut or vegetable oil
6 to 6½	cups small broccoli flowerets
1	small red onion, coarsely chopped

Stir together the sugar, water, vinegar, soy sauce, cornstarch, and spices in a small bowl or cup until well blended and smooth. Set aside.

In a large skillet over high heat, heat the oil to hot but not smoking. Add the broccoli and onion and cook, stirring constantly, for 1½ minutes. Lower the heat to medium and add 2 tablespoons of water to the skillet. Cook, stirring, for about 1½ minutes longer, or until most of the liquid has evaporated from the pan and the broccoli is almost tender. Stir the sauce briefly and add it to the skillet. Cook, stirring, for 1 to 1½ minutes longer, or until the sauce thickens slightly and is clear and the broccoli is crisp-tender.

Makes about 4 servings.

BROCCOLI ITALIAN STYLE

(side dish)

2	tablespoons olive oil
1	large garlic clove, minced
6	cups small broccoli flowerets
⅓	cup coarsely chopped sweet red pepper (optional)
¼ to ⅓	cup dry white wine (approximately)
¼	teaspoon salt
⅛	teaspoon black pepper, preferably freshly ground
1	tablespoon grated Parmesan cheese

Heat the olive oil in a large skillet or sauté pan over high heat until hot but not smoking. Add the garlic and cook, stirring, for 30 seconds. Add the broccoli and red pepper (if used) and cook, stirring, for 1 minute longer. Stir in ¼ cup of the wine and the salt and pepper. Lower the heat to medium. Cook the vegetables, stirring, for 4 to 6 minutes longer, or until the broccoli is crisp-tender and most of the wine has evaporated from the pan. (If the pan begins to boil dry before the broccoli is cooked through, add a tablespoon or two more wine.) Transfer the mixture, along with any pan liquid, to a serving dish. Sprinkle the mixture with the Parmesan cheese.

Makes 4 to 6 servings.

HERBED BROCCOLI WITH SESAME SEEDS

(side dish)

3	medium-sized fresh broccoli stalks, including stems
1½	tablespoons vegetable or sesame oil
1	large onion, finely chopped
2	garlic cloves, minced
2 to 3	tablespoons sesame seeds
1	teaspoon dried thyme leaves
½	teaspoon dried basil leaves
1	cup sliced fresh mushrooms (optional)
2	tablespoons water
⅛	teaspoon salt (optional, or to taste)

Cut the broccoli tops into small flowerets. Trim and discard ½ inch from the bottom of each stem. Peel the woody covering from the thick part of the stems and discard it; then cut the stems into thin slices. Set aside the flowerets and stem pieces.

In a large skillet over medium-high heat, heat the oil; then cook the onion and garlic until they are tender but not browned. Stir in the sesame seeds and sauté for about 1 minute longer. Add the thyme, basil, mushrooms (if used), and the reserved broccoli flowerets and stem pieces. Cook for about 2 minutes while stirring constantly; then add the water and cover the skillet tightly. Cook the broccoli for about 5 to 8 minutes, or just until it is crisp-tender and brightly colored. If desired, season with salt. Serve immediately.

Makes about 6 servings.

EASY CRUMB-TOPPED BROCCOLI BAKE

(side dish)

Convenient, attractive, and tasty, this broccoli casserole makes a festive addition to any meal. Because it can be assembled completely in advance, it is perfect for entertaining.

1½	tablespoons butter or margarine
2	tablespoons enriched all-purpose or unbleached white flour
1	teaspoon prepared mustard, preferably Dijon or Dijon-style
1	cup chicken broth or bouillon (reconstituted from cubes or granules)
2	teaspoons instant minced onions
⅛	teaspoon black pepper, preferably freshly ground
½	cup whole milk
1	20-ounce package loose-pack frozen broccoli cuts, thawed and drained

Crumb Topping

1¾	cups commercial herb-seasoned stuffing
⅓	cup chicken broth or bouillon (reconstituted from cubes or granules)
1½	tablespoons cold butter or margarine

Melt the butter in a medium-sized saucepan over medium-high heat. Gradually stir in the flour until it is well blended. Cook, stirring, for 2 minutes. Gradually stir in the mustard and then the chicken broth, until the mixture is well mixed and smooth. Add the onions and pepper and bring the mixture to a simmer. Cook, stirring occasionally, for 1½ to 2 minutes, or until thickened and smooth. Stir in the milk and heat until the sauce mixture is hot but not boiling. Set aside.

Spread the broccoli in a 9- or 10-inch-square baking dish. Spoon the sauce evenly over the broccoli. Stir together the stuffing mix and chicken broth until the crumb mixture is evenly moistened. Sprinkle evenly over the broccoli. Dot the top with the butter.

Cover the casserole and bake for 30 minutes in a preheated 350-degree oven. Remove the cover and continue baking for 15 to 20 minutes, or until the casserole is bubbly and the top is slightly crisp.

Makes 5 to 7 servings.

BROCCOLI-RICE CASSEROLE

(side dish or light main dish)

2 cups *cooked* white or brown rice
2 medium-sized fresh broccoli stalks, including stems
2 tablespoons butter or margarine
1 medium-sized onion, finely chopped
1 garlic clove, minced
½ teaspoon dried thyme leaves
½ teaspoon dried oregano leaves
¼ teaspoon dried dillweed
¼ cup finely chopped fresh parsley leaves
1 medium-sized sweet green pepper, thinly sliced
2 tablespoons water
¼ cup unsalted cashews or other nuts, or unsalted sunflower seeds
½ cup plain lowfat yogurt
4 ounces Swiss or similar cheese, grated (1 cup packed)

Press the rice into the bottom of a greased or nonstick spray-coated 8-inch-square or 9-inch-round baking dish. Set aside.

Cut the broccoli tops into small flowerets. Trim and discard ½ inch from the bottom of each stem. Peel the woody covering from the thick part of the stems and discard it; then cut the stems into thin slices. Set aside the flowerets and stem slices.

In a large skillet over medium-high heat, melt the butter; then cook the onion and garlic until they are tender but not browned. Stir in the herbs, green pepper, and the reserved broccoli flowerets and stem slices. Add the water to the skillet, cover tightly, and braise the vegetables for about 5 minutes, or until the broccoli is brightly colored and crisp-tender. Remove from the heat, and stir in the nuts.

Spread the vegetable mixture over the rice. Spoon the yogurt on top in dollops. Then sprinkle the grated cheese over the vegetables and yogurt. Bake the casserole, uncovered, in a preheated 350-degree oven for about 20 minutes, or until the cheese is melted.

Makes 4 to 6 side-dish servings or 2 to 3 main-dish servings.

PASTA PRIMAVERA

(side dish or light main dish)

This is often made with lots of olive oil and/or heavy cream. Our version cuts way down on fat, but still has a rich flavor.

1 to 2	tablespoons butter or margarine
1	large onion, finely chopped
1	medium-sized leek (white and pale green parts only), well washed and thinly sliced
4	cups broccoli flowerets and thinly sliced peeled stems
1	cup thinly sliced carrots
¾	cup thinly sliced celery-cabbage (or celery stalks)
1	sweet red pepper, cut into ¾-inch squares (if unavailable, substitute sweet green pepper)
¼	cup water
8	ounces uncooked spaghetti
1	15-ounce container part-skim ricotta cheese (if unavailable, substitute regular ricotta)
1	tablespoon cornstarch
¾	cup skim or lowfat milk
¼	cup grated Parmesan cheese
½	teaspoon dried basil leaves
½	teaspoon dried marjoram leaves
½	teaspoon dried thyme leaves
¼	teaspoon salt
⅛	teaspoon black pepper, preferably freshly ground

In a very large deep skillet, over medium-high heat, melt the butter; then sauté the onion and leek until they are tender but not browned. Stir in the broccoli, carrots, celery-cabbage, and sweet pepper. Add the water, cover tightly, and steam until the vegetables are tender, about 10 minutes.

Meanwhile, break the spaghetti in half lengthwise; then cook it according to the package directions. Drain the spaghetti well and set it aside.

In a medium-sized bowl, combine the ricotta and cornstarch. Then mix in the milk, Parmesan cheese, herbs, and seasonings. When the vegetables are tender, stir the ricotta mixture into the skillet. Cook over medium heat, stirring constantly, until the sauce is heated through. Gently stir in the cooked spaghetti. Serve hot.

Makes about 8 side-dish servings or 4 main-dish servings.

Variation

NOT-PASTA PRIMAVERA!

In this version of the dish, spaghetti squash strands are substituted for the pasta. Cook an approximately 3-pound spaghetti squash until it is just done and not mushy (see the basic directions with the recipe for Spaghetti Squash with Herbs and Cheese, page 259). Use a fork to carefully pull the strands from the squash. This should yield about 4 cups of cooked squash. Use that in the above recipe in place of the cooked pasta.

VEGETABLE-CHEESE PUFF

(light main dish)

1   tablespoon butter or margarine
2   cups mixed small broccoli and cauliflower flowerets
1   small onion, chopped
1   garlic clove, minced
¼   cup lowfat milk
½   cup lowfat cottage cheese
3   large eggs, separated
½   teaspoon salt
⅛   teaspoon black pepper, preferably freshly ground
½   teaspoon dried basil leaves
¼   teaspoon powdered mustard
2   cups fresh whole wheat or white bread cubes
5   ounces mild Cheddar or lowfat Longhorn cheese, grated or shredded (1¼ cups packed), divided

In a large skillet, melt the butter over medium-high heat. Add the broccoli and cauliflower, onion, and garlic. Cook, stirring constantly, until the onion is tender. Set aside.

In a nonstick spray-coated 1-quart casserole, combine the milk, cottage cheese, and slightly beaten egg yolks. Add the salt, pepper, basil, mustard, and bread cubes. Mix well. Stir in 1 cup of the cheese. Then fold in the cooked vegetables.

Beat the egg whites until just foamy. Fold into the vegetable-bread mixture, combining well. Sprinkle the remaining cheese over the top of the casserole. Bake in a preheated 350-degree oven for about 35 minutes, or until the cheese puff is set.

Makes 3 to 4 servings.

CHICKEN AND BROCCOLI SALAD

(main dish salad)

The unusual combination of flavors and ingredients makes this salad a nice change of pace from ordinary chicken salad.

3	cups small broccoli flowerets
¼	cup plain lowfat yogurt
½	cup mayonnaise
1	teaspoon dried dillweed
½	teaspoon celery salt
	Pinch of black pepper, preferably freshly ground
1½ to 2	cups ¾-inch cooked chicken breast meat cubes
½	cup walnuts, coarsely chopped
½	cup golden raisins (if unavailable, use dark raisins)
1	small apple, cored and cut into ¼-inch cubes
1	celery stalk, coarsely chopped

In a medium-sized pot, combine the broccoli flowerets with ¾ cup water. Cover and bring to a boil. Lower the heat and simmer the broccoli for 3 to 4 minutes, or until it is just slightly tender. Drain very well in a colander.

Meanwhile, in a medium-sized bowl, combine the yogurt, mayonnaise, dillweed, celery salt, and black pepper. Stir well. Add the broccoli, chicken, walnuts, raisins, apple, and celery. Stir to mix well. Cover and refrigerate for several hours or overnight so that the flavors can blend.

Makes about 4 servings.

QUICK TUNA-VEGETABLE SKILLET DINNER

(main dish)

This easy dish is a good way to use up small amounts of different frozen vegetables.

1   tablespoon vegetable oil
1   medium-sized onion, finely chopped
1   garlic clove, minced
1   cup thinly sliced celery
2   cups small broccoli flowerets and stem slices
2   cups fresh or frozen mixed vegetables, such as thinly sliced or diced carrots, cut string beans, corn kernels, peas, etc.
½   cup water
3   cups *cooked* white or brown rice, or *cooked* bulgur wheat
2   6½-ounce cans water-packed tuna, drained and coarsely flaked
3   tablespoons soy sauce
1   cup drained canned chick-peas (optional)

In a very large deep skillet over medium-high heat, heat the oil; then cook the onion, garlic, and celery, stirring, until they are tender but not browned. Add the broccoli, mixed vegetables, and water and bring to a boil. Lower the heat, cover the pan tightly, and simmer the vegetables for about 5 to 10 minutes, or until they are just crisp-tender. Stir in the cooked rice, tuna, soy sauce, and chick-peas (if used). Heat, stirring gently, until the ingredients are completely warmed through.

Makes about 4 servings.

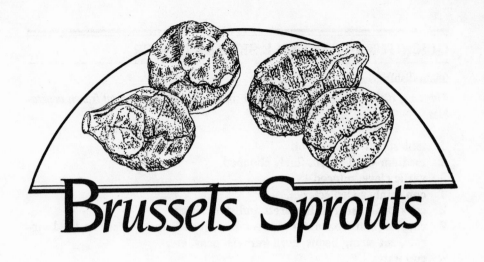

Brussels Sprouts

Brussels sprouts look like compact, tightly furled miniature heads of cabbage, and little wonder, since this is in fact what they are! These pale green nubbins are the products of a tall-stemmed cabbage variety that sprouts numerous small heads along its stem instead of a single large one at the end of the stalk. The edible buds grow in a spiral pattern from the bottom of the plant upward, and as these mature they are simply plucked off the stem. Although the Brussels sprouts plant has a two-year life cycle, it only bears its little cabbages the first year; flowers and seeds are produced the second.

Some sources suggest that Brussels sprouts have been grown in Belgium for many centuries, but most experts feel that they are a fairly modern vegetable, having evolved from a variety of Savoy cabbage in the 1600s or 1700s. (Interestingly, small buds will sometimes also form along the stem of regular cabbage if the head is cut off.) In any case, by the 1800s, Brussels sprouts were popular in Belgium, and were being cultivated in France and England as well. The "thousand-headed" cabbages, as one nineteenth-century American horticulturalist described them, did not become widely known or of much commercial importance in the United States until the 1920s. The market again expanded somewhat with the advent of the frozen food industry in the 1940s.

Today, Brussels sprouts are particularly valued as a cool weather crop. They tolerate low temperatures better than almost all other cabbage varieties, and produce their best "heads" during a fall with sunny days and crisp nights.

Availability: Fresh Brussels sprouts can be found all year round, although the supply is largest from September to February and tapers off from April to August. Frozen Brussels sprouts are also readily available throughout the year and make a reasonably good substitute for fresh.

Choosing the Best: The best Brussels sprouts are firm, smooth, tightly headed and have a pale green, waxy sheen. The smaller, less mature buds are usually the tenderest and have the most delicate flavor. Avoid sprouts that are puffy or yellowed or that have loose, shriveled leaves, as these are past their prime and are likely to have a strong taste.

Nutritional Value: In addition to providing a large amount of vitamin C and some vitamin A, Brussels sprouts contain significant amounts of thiamine, iron, potassium, and phosphorus.

Storage: Brussels sprouts are highly perishable and should be placed in plastic bags and quickly refrigerated to prevent the leaves from yellowing. Moreover, they should not be stored longer than a day or two or they may develop a strong cabbage-like taste.

Preparation and Basic Cooking: Brussels sprouts are most often cooked whole, and, in this case, very little preparation is required. Simply rinse them off; then trim away a thin slice at the bottom of each head and cut an X into the core to encourage even cooking. Also pull or trim away any wilted or yellowed outer leaves. Simmer the sprouts in a small amount of water until they are just barely tender; tiny ones may require as little as 6 to 8 minutes and large ones may take 12 to 15. The sprouts may also be steamed for about the same length of time in a steamer basket. However, this method tends to dull their attractive green color somewhat. This loss of color can be minimized by lifting the pot lid and allowing the steam to escape several times during cooking.

Another possibility is to slice the heads in half lengthwise. They can then be quickly braised in a little chicken broth or stir-fried in a mixture of butter and vegetable oil until they are crunchy-crisp.

Always serve Brussels sprouts promptly, as they can develop an overly strong taste if they stand too long.

Simple Serving Suggestions: Cooked whole Brussels sprouts are good dressed with just a bit of butter and salt and pepper. A dash of lemon juice or a sprinkling of toasted pecans or walnuts will also add an appealing note.

HERBED BRUSSELS SPROUT SOUP

Even Brussels sprouts "haters" can sometimes be won over by this tasty soup.

2	tablespoons butter or margarine
5 to 6	scallions, including 1 inch of the green tops, coarsely chopped
2	medium-sized celery stalks, coarsely chopped
3½	tablespoons enriched all-purpose or unbleached white flour
3½	cups chicken broth, preferably homemade, or bouillon (reconstituted from cubes or granules)
2	teaspoons chopped fresh chives (or 1 teaspoon dried chopped chives)
½	teaspoon salt (decrease to ¼ teaspoon or omit if commercial chicken broth is used)
⅛ to ¼	teaspoon white pepper, preferably freshly ground
	Pinch of dried tarragon leaves
	Pinch of dried thyme leaves
¾	pound fresh Brussels sprouts (or 1 10-ounce package frozen Brussels sprouts, thawed), cut in half lengthwise
⅔	cup milk (approximately)

Melt the butter in a 3- to 4-quart saucepan or pot over medium-high heat. Add the scallions and celery and cook, stirring, for 4 to 5 minutes, or until the scallions are tender but not browned. Stir in the flour until well blended and cook for 2 minutes longer. Gradually add the chicken broth, stirring until the mixture is well blended and smooth. Stir in all remaining ingredients, *except* the milk, and bring the mixture to a boil. Lower the heat and simmer the soup, uncovered, for 8 to 12 minutes, or until the Brussels sprouts are just barely tender (simmer 4 to 7 minutes if frozen sprouts are used). In batches, if necessary, transfer the soup to a blender and blend on medium speed until the vegetables are completely puréed. (If a slightly chunky soup is preferred, about ½ cup Brussels sprout halves may be reserved in a small bowl and then re-combined with the puréed mixture.)

Return the puréed mixture (and Brussels sprouts halves, if reserved) to the saucepan. Stir in ⅔ cup milk, or enough to yield a substantial, yet not thick, consistency. Heat the soup to piping hot and serve.

Makes 4 to 6 servings.

MARINATED VEGETABLES

Vegetables
½ cup water
1 10-ounce package frozen Brussels sprouts
3 medium-sized carrots, thinly sliced
3 cups small cauliflower flowerets
1 16-ounce can cut green beans, well drained
1 15- to 16-ounce can chick-peas, well drained

Marinade
⅓ cup apple cider vinegar
⅓ cup vegetable oil
2 tablespoons drained chopped pimiento
2 tablespoons chopped fresh parsley leaves
2 tablespoons chopped scallions (green tops only)
1 tablespoon sugar
½ teaspoon dried basil leaves
¼ teaspoon powdered mustard
¼ teaspoon salt
⅛ teaspoon black pepper, preferably freshly ground

In a medium-sized saucepan, bring the water to a boil over high heat. Add the Brussels sprouts. Lay the carrots and cauliflower over the Brussels sprouts. Lower the heat and simmer, covered, for about 6 minutes, or until the Brussels sprouts are tender and the carrots are cooked but still crisp-tender. Drain the vegetables well in a colander and cool slightly. Cut any large Brussels sprouts in half.

In a large bowl, combine the cooked vegetables with the canned green beans and chick-peas.

In a small bowl or jar, combine all the marinade ingredients. Stir or shake to mix well. Add to the vegetables and toss to coat them. Refrigerate, covered, for several hours to allow the flavors to blend. Toss before serving.

Makes about 8 servings.

BRUSSELS SPROUTS WITH TARRAGON SAUCE

(side dish)

Although the combination of ingredients in the light sauce may seem a bit unusual, it has a very appealing taste. It would also be suitable with other steamed green vegetables, such as green beans, broccoli, asparagus, and peas.

1¼   pounds fresh Brussels sprouts, trimmed*

Sauce
** 1   tablespoon butter or margarine**
1½   tablespoons enriched all-purpose or unbleached white flour
** 1   cup chicken broth or bouillon (reconstituted from cubes or granules)**
** ½   teaspoon dried tarragon leaves**
** 1   teaspoon prepared Dijon-style mustard**
** 1   tablespoon apple cider vinegar**

In a steamer basket set above some simmering water, or in a medium-sized saucepan containing about 1 inch of water, steam the Brussels sprouts until they are just tender, but not at all mushy, about 8 to 15 minutes.

Meanwhile, prepare the sauce. In a small saucepan over medium-high heat, melt the butter. Add the flour and stir constantly, preferably with a wire whisk, for 1 minute. Then gradually mix in the broth, tarragon, mustard, and vinegar, and continue stirring until the sauce thickens and comes to a boil. Lower the heat and simmer the sauce for 1 minute. If necessary, keep the sauce warm over very low heat, stirring often, until it is needed.

When the Brussels sprouts are tender, immediately remove them from the heat and drain off any cooking liquid. Transfer the Brussels sprouts to a serving dish and top with the sauce. Stir gently so that all the Brussels sprouts are coated with some of the sauce.

Makes 5 to 6 servings.

*Fourteen to 16 ounces of loose-pack frozen Brussels spouts may be substituted for the fresh ones. Follow the package directions for basic preparation.

BRUSSELS SPROUT AND CAULIFLOWER STIR-FRY

(side dish)

Quick, colorful, and deliciously crunchy and fresh-tasting, this unusual stir-fry is also extremely healthful.

Sauce
½   cup chicken broth or bouillon (reconstituted from cubes or granules)
1   tablespoon soy sauce
1   teaspoon lemon juice
2   teaspoons cornstarch
⅛   teaspoon black pepper, preferably freshly ground

Vegetables
2   tablespoons peanut or vegetable oil
1   garlic clove, minced
1   small red onion, coarsely chopped
¾   pound small fresh Brussels sprouts (about 25 to 30), trimmed and cut in half lengthwise
2½   cups medium-sized cauliflower flowerets
1   large celery stalk, coarsely chopped

Combine all the sauce ingredients in a small bowl, stirring until the cornstarch is incorporated and the mixture is smooth. Set aside.

In a large skillet over high heat, heat the oil until hot but not smoking. Stir in the garlic, onion, Brussels sprouts, cauliflower, and celery. Cook, stirring constantly, for 2 minutes; be sure the vegetables do not brown. Lower the heat to medium-high. Briefly stir the reserved sauce mixture and add it to the skillet. Cook, continuing to stir, for 5 to 7 minutes longer, or until the Brussels sprouts and cauliflower are just barely crisp-tender.

Makes 5 to 6 servings.

SAUCY BRUSSELS SPROUTS

(side dish)

1   10-ounce package frozen Brussels sprouts
1   tablespoon butter or margarine
1   tablespoon enriched all-purpose or unbleached white flour
¾   cup whole or lowfat milk
⅛   teaspoon ground cloves
⅛   teaspoon ground thyme
⅛   teaspoon powdered mustard
¼   teaspoon celery salt
⅛   teaspoon onion powder
    Pinch of black pepper, preferably freshly ground

In a small saucepan, cook the Brussels sprouts according to the package directions. Drain them in a colander.

While the vegetables are cooking, melt the butter in a small saucepan over medium-low heat. Blend in the flour with a spoon. Cook, stirring, until the mixture is smooth. Gradually add the milk, stirring with a wire whisk. Raise the heat to medium. Cook, stirring, until the sauce thickens and boils. Stir in the cloves, thyme, mustard, celery salt, onion powder, and pepper. Blend well. Add the Brussels sprouts to the pot and stir to coat with the sauce. Serve at once.

Makes 3 to 4 servings.

BRUSSELS SPROUTS WITH LEMON AND PARMESAN CHEESE

(side dish)

1¼   pounds fresh Brussels sprouts, trimmed (about 25 to 30 medium sized)
½   cup chicken broth or bouillon (reconstituted from cubes or granules)
⅛   teaspoon black pepper, preferably freshly ground
1   tablespoon butter or margarine
1½   teaspoons lemon juice, preferably fresh
2   teaspoons grated Parmesan cheese

Combine the Brussels sprouts, chicken broth, and pepper in a medium-sized sauce-pan and bring to a boil over medium-high heat. Lower the heat and simmer the sprouts, uncovered, for 8 to 12 minutes, or until they are just tender and most of the liquid has evaporated. (If necessary, add a bit of water to the pan to prevent it from boiling dry during cooking.) Add the butter and lemon juice and cook, stirring, for about 30 seconds longer, or until the butter melts. Transfer the Brussels sprouts and any liquid remaining in the pan to a serving dish. Sprinkle the sprouts with the grated Parmesan cheese.

Makes 4 to 5 servings.

Cabbage

According to Greek myth, cabbages first came about when the god of wine, Dionysus, caught the ancient king Lycurgus destroying some grapevines. The misdoer was tied to a grapevine to await severe punishment, at which point he cried. As his many tears fell to the ground, they became cabbages.

Historically, however, we cannot be sure about the actual origin of cabbage. One of the oldest vegetables known, it has been cultivated for at least 4,000 years.

The cabbage family is extremely diverse and versatile. In addition to the many different types and colors of head cabbages, it also includes Brussels sprouts, broccoli, cauliflower, collards, kale, and kohlrabi. Incredibly, these varieties did not develop through mutation or hybridization, but simply by emphasizing various parts of the original plant that already existed.

Celery-cabbage, Napa cabbage, and bok choy are not members of the same species as the cabbage family; however, they are closely related to it. Although these three are most often featured in Oriental cuisine, they have become increasingly popular for general use in recent years.

Descriptions of the major types of cabbages follow:

Bok choy has branching white to greenish-white stalks with edible dark green leaves at the ends. The stalks have a mild taste and crisp texture. Bok choy is sometimes called by the misnomer "Chinese chard."

Celery-cabbage has broad, white, celery-like stalks, which are actually the tender, edible ribs of large, pale green to yellowish-green leaves. The heads are quite long and narrow and usually have more rib than leaf. This is sometimes called "Chinese celery-cabbage."

Green or white cabbage is probably the most commonly available type, and it is what people generally think of as "cabbage." The heads are firm, compact, and spherical, with smooth, very pale green leaves that tightly overlap one another.

Napa cabbage has pale green, oblong heads. The leaves look and taste like a cross between celery-cabbage and Savoy cabbage, and may even resemble romaine lettuce. This is sometimes called "Chinese cabbage" or "Nappa."

Red cabbage is actually reddish-purple in color. It is very similar to green cabbage in shape and texture, but has a slightly sweeter taste. Raw and shredded, it is often added to tossed salads for color and textural contrast.

Savoy cabbage is known for its puckered, crinkly, flexible green leaves, which are loosely packed in a spherical-to-ovoid head. Its flavor is usually milder than green cabbage.

Availability: Green and red cabbages are usually in plentiful supply the year round. The availability of the other varieties depends somewhat on demand and location.

Choosing the Best: In all cabbages, look for well-formed, compact heads that do not have wilted or discolored outer leaves. Green and red cabbages should be solid and heavy for their size, with very tight leaves. Savoy and Napa cabbages will be looser and lighter in weight for their size than green cabbage, and have crisper, crinkly leaves. Celery-cabbage and bok choy should have white to greenish-white stalks with no discoloration or dark spots. In general, a fresh scent is indicative of a fresh (rather than stored) cabbage.

Nutritional Value: Cabbage is rich in vitamin C and has fair amounts of vitamin A and thiamine, as well as several minerals. Bok choy is particularly high in calcium. Cabbage is also a good source of non-nutritive fiber.
(In addition, members of the cabbage family, also known as cruciferous vegetables, may be able to help the body fight some forms of cancer.)

Storage: The firm green and red cabbages will keep at least 7 to 10 days loosely wrapped in a plastic bag in the refrigerator. The softer ones have a shorter storage time. Humidity is particularly important with the latter to keep their leaves from wilting. In general, the sooner cabbage is used, the less likely it is to have an "off" odor when cooked.

Preparation and Basic Cooking: For best nutritional value, do not cut or shred cabbage until just before using it. After washing the cabbage, remove and discard any discolored or shriveled outer leaves, and core the cabbage if desired.

To cook cabbage, thinly slice or shred it, or cut round-head cabbages into portion-size wedges with a small part of the core intact (to hold the leaves together). Then, steam or braise the cabbage in a small amount of liquid just until it is tender, about 10 to 15 minutes depending on the size of the pieces. If desired, the wedges may be boiled, which will take a few minutes less than steaming. Do not overcook cabbage, as this gives it a very limp, unappealing texture, and may produce the strong odor often associated with this vegetable.

In addition, celery-cabbage or bok choy may be quickly stir-fried in a small amount of oil with other vegetables and perhaps meat or poultry.

Simple Serving Suggestions: Shredded or thinly sliced raw cabbage of all types is a great addition to tossed salad, adding texture, color, and flavor. With a little dressing stirred in, cabbage can be turned into coleslaw. Cooked cabbage wedges are tasty with some butter. When cooking red cabbage, add a little vinegar or lemon juice to help retain the bright color. Parboiled whole leaves of green or Savoy cabbage may be rolled around a meat and/or rice mixture and braised in a sauce until the filling is cooked through.

TANGY CABBAGE AND POTATO SOUP

2   tablespoons olive oil
2   cups finely chopped onion
1   stalk celery, thinly sliced
2   large carrots, grated or shredded
½   cup diced peeled rutabaga
2   garlic cloves, minced
4½   cups water
2   cups vegetable or beef bouillon (reconstituted from cubes or granules) or vegetable stock (page 86)
2   tablespoons apple cider vinegar
1½   teaspoons sugar
½   teaspoon powdered mustard
½   teaspoon dried marjoram leaves
¼   teaspoon dried thyme leaves
1   teaspoon salt
¼   teaspoon black pepper (or more to taste), preferably freshly ground
2   cups shredded green cabbage
1   cup finely shredded peeled potato

Heat the oil in a Dutch oven over medium-high heat. Add the onion and celery and cook, stirring, until the onion is soft. Add the rest of the ingredients, *except* the cabbage and the potato, and bring the soup to a boil. Cook for 30 minutes. Add the cabbage and potato. Raise the heat and make sure the soup boils again. Then lower the heat and simmer the soup for an additional 25 to 30 minutes, or until the vegetables are tender and the potato has thickened the liquid slightly.

Makes 6 to 8 servings.

SWEET AND SOUR GROUND BEEF AND VEGETABLE SOUP

1	pound lean ground beef
8	cups water
1	8-ounce can tomato sauce
1	small turnip, peeled and grated or finely chopped
1	celery stalk, finely chopped
1	medium-sized carrot, grated or shredded
1	large onion, grated or finely chopped
1	small zucchini, ends trimmed, grated or finely chopped
2	garlic cloves, minced
2	tablespoons pearl barley
1½	tablespoons sugar
1½	tablespoons apple cider vinegar
1	teaspoon dried basil leaves
1	teaspoon prepared horseradish
2 to 3	bay leaves
¾	teaspoon powdered mustard
¼	teaspoon ground thyme
3 to 4	drops Tabasco sauce
1½	teaspoons salt
	Generous ¼ teaspoon black pepper, preferably freshly ground
2	medium-sized potatoes, peeled and cut into ½-inch cubes
1	large carrot, thinly sliced
½	cup fresh or loose-pack frozen corn kernels
3	cups coarsely chopped green cabbage

In a large, heavy pot over medium heat, brown the ground beef, stirring and breaking up the meat with a spoon. Drain off and discard all fat. Add the remaining ingredients, *except* the potatoes, sliced carrot, corn, and cabbage. Bring to a boil. Lower the heat and simmer for 1 hour, stirring occasionally.

Add the remaining vegetables. Raise the heat and bring the soup to a boil. Then lower the heat again, cover, and simmer, stirring occasionally, for about 30 minutes longer, or until the potatoes and carrots are tender.

Makes about 8 servings.

RAINBOW CABBAGE SLAW

Colorful, tasty, and healthful, this makes a great picnic or buffet dish. It is less per-ishable than many slaws and, thus, very convenient to "take along" in hot weather.

1	2-pound head green cabbage, shredded
2	large carrots, grated or shredded
¼	cup chopped red onion
1	small sweet green pepper, finely chopped
1	small sweet red pepper, finely chopped
5 to 6	medium-sized red radishes

Dressing

⅔	cup apple cider vinegar
¼	cup sugar
¼	cup vegetable oil
¾	teaspoon celery seeds
½	teaspoon powdered mustard
¼	teaspoon black pepper, preferably freshly ground
¼	teaspoon salt

Combine all the vegetables in a large, noncorrosive bowl and toss until well mixed.

Combine all the dressing ingredients in a jar with a tight-fitting lid or a cruet and shake vigorously until the sugar dissolves and the ingredients are well blended. Add the dressing to the slaw and stir until thoroughly mixed. Cover and refrigerate the slaw for at least 30 minutes to allow the flavors to blend. It keeps well, refrigerated, for 3 or 4 days.

Makes 8 to 10 servings.

CALICO COLESLAW WITH COTTAGE DRESSING

The red cabbage in this "slaw" turns the dressing a pretty pink color.

4   cups finely shredded green cabbage (about 1 pound)
4   cups finely shredded red cabbage (about 1 pound)
1   small onion, grated
1   cup regular or lowfat small curd cottage cheese
¾   cup commercial buttermilk
⅓   cup mayonnaise or mayonnaise-like "dressing"
¼   cup apple cider vinegar
4   teaspoons sugar
½   teaspoon powdered mustard
½   teaspoon celery seeds
½   teaspoon caraway seeds (optional)

In a large bowl, combine the green and red shredded cabbage with the onion.

In a medium-sized bowl, stir together the remaining ingredients. Pour this over the cabbage mixture, and toss well.

Cover and refrigerate the coleslaw for several hours or, preferably, overnight, to give the flavors a chance to blend. Toss again before serving.

Makes about 8 servings.

CABBAGE WITH TARRAGON

(side dish)

This is quick and easy, yet very tasty.

1   tablespoon water
1   tablespoon dry white wine or chicken broth
1½   tablespoons tarragon vinegar or apple cider vinegar
1   tablespoon sugar
2   tablespoons chopped fresh chives (or 1 tablespoon dried chopped chives)
½   teaspoon dried tarragon leaves
⅛   teaspoon salt
     Scant ⅛ teaspoon black pepper, preferably freshly ground
1   tablespoon peanut or vegetable oil
1   large onion, coarsely shredded
8   cups coarsely shredded green cabbage

Combine the water, wine, vinegar, sugar, chives, tarragon, salt, and pepper in a small bowl or cup. Stir until the sugar dissolves; set the seasoning mixture aside.

Heat the oil in a very large skillet over high heat. Add the onion and cabbage and cook, stirring constantly, for 2 minutes, or until the onion is slightly soft. Add the reserved seasoning mixture to the pan; then cook, stirring constantly, for 6 to 7 minutes, or until most of the excess liquid evaporates and the cabbage has cooked down but is still slightly crisp.

Makes 4 servings.

BRAISED CABBAGE AND SWEET PEPPERS

(side dish)

1	tablespoon vegetable oil
½	tablespoon butter or margarine
3 to 4	scallions, including green tops, coarsely chopped
1	small carrot, shredded or finely chopped
1	small sweet green pepper, coarsely chopped
1	small sweet red pepper, coarsely chopped (if unavailable, substitute another sweet green pepper)
1	1¼- to 1½-pound head green cabbage, coarsely shredded
½	cup beef broth or bouillon (reconstituted from cubes or granules)
1½	tablespoons ketchup
¾	teaspoon paprika
⅛	teaspoon salt
⅛	teaspoon black pepper, preferably freshly ground

Heat the oil and butter in a very large skillet or sauté pan (at least 12 inches in diameter) over high heat until hot but not smoking. Add the scallions, carrot, and sweet peppers and cook, stirring, for 1½ minutes. Stir in all the remaining ingredients, tossing until well blended. (The cabbage will fill the pan at first, but then it will gradually decrease in volume.) Lower the heat to medium and cook, stirring, for 5 to 7 minutes longer, or until the cabbage is cooked through but still slightly crisp.

Makes 4 to 6 servings.

HERBED CABBAGE AND POTATO CASSEROLE

(light main dish or side dish)

6 to 8	small russet potatoes, scrubbed but not peeled
1	tablespoon butter or margarine
1	medium-sized onion, finely chopped
1	garlic clove, finely minced
4	cups packed, shredded cabbage (white, Savoy, Chinese cabbage, red cabbage, etc.)
2	cups small curd lowfat cottage cheese
1	large egg, lightly beaten
¼	teaspoon dried thyme leaves (or to taste)
¼	teaspoon caraway seeds (or to taste)
¼	teaspoon salt
	Scant ⅛ teaspoon black pepper, preferably freshly ground
2 to 4	tablespoons grated cheese, such as Swiss or Muenster

Cut the potatoes into 1-inch pieces, then put them in a medium-sized saucepan with about 1 inch of water. Bring to a boil over high heat. Cover tightly and lower the heat until the water is just simmering. Cook the potatoes for about 12 to 15 minutes, or until they are tender.

Meanwhile, melt the butter in a large skillet over medium-high heat and cook the onion and garlic, stirring, until they are tender but not browned. Add the cabbage and cook until it is wilted. Set aside.

As soon as the potatoes are done, drain them very well; then use a fork to coarsely mash them right in the saucepan. Stir in the cottage cheese and egg. Add the potato-cheese mixture to the cooked cabbage mixture; then stir in the thyme, caraway seeds, salt, and pepper.

Transfer the vegetable mixture to a greased or nonstick spray-coated 10-inch-round quiche dish or a 9-inch-square pan. Sprinkle the grated cheese on top. Bake in a preheated 350-degree oven for about 30 minutes, or until the cheese is melted and the casserole is set. Serve cut into wedges or squares.

Makes about 3 light main-dish servings or about 6 side-dish servings.

SAVOY CABBAGE BAKE

(side dish)

This is a colorful, savory, and unusual way to serve one of our most healthful vegetables.

Sauce
1 tablespoon butter or margarine
2 tablespoons enriched all-purpose or unbleached white flour
¾ cup whole milk
1 teaspoon instant minced onions
½ teaspoon salt
¼ cup commercial sour cream

Casserole
1 medium-sized head Savoy cabbage (about 1¼ pounds), cored and tough outer leaves removed
1 large tomato, peeled, seeded, and finely chopped
Salt and black pepper to taste
1 tablespoon imitation bacon bits
2 ounces Monterey Jack cheese, shredded (½ cup packed)

To prepare the sauce, melt the butter in a small heavy saucepan over medium-high heat. Stir in the flour until well blended. Cook the mixture, stirring, for 2 minutes. Gradually add the milk to the pan, stirring vigorously until the mixture is smooth and well blended. Stir in all remaining sauce ingredients, *except* the sour cream, and bring the mixture to a boil. Boil for 1 minute, stirring; then remove from the heat and stir in the sour cream. Set the sauce aside.

To assemble the casserole, cut the cabbage into 6 equal wedges. Lay the wedges in a colander. Lightly blanch the wedges by pouring about 2 quarts of boiling water over them. Drain the wedges and pat them dry with paper towels. Transfer the wedges to an 8-inch-square by 3-inch-deep baking dish (or similar-sized casserole). Sprinkle the chopped tomato over the wedges. Sprinkle the cabbage and the tomatoes with salt and pepper. Spoon the reserved sauce over the cabbage wedges, dividing it equally among them. Top each wedge with some of the bacon bits and then some of the cheese. Cover the casserole. Bake in a preheated 350-degree oven for 30 to 35 minutes, or until the sauce and cheese are bubbly and the cabbage is tender when the thickest part is pierced with a fork.

Makes 6 servings.

CELERY-CABBAGE AND PORK SKILLET

(main dish)

- ¾ pound lean boneless pork loin or shoulder, trimmed and cut into ½-inch cubes
- 2 tablespoons soy sauce
- ½ teaspoon minced fresh gingerroot
- 3 tablespoons dry sherry or dry white wine
- 1 tablespoon packed light or dark brown sugar
- 1 tablespoon ketchup
- 1 to 2 drops Tabasco sauce
- 2¾ teaspoons cornstarch
- 1 1⅓- to 1½-pound head celery-cabbage
- 7 to 8 scallions
- 2 tablespoons peanut or vegetable oil
- 1 large garlic clove, minced

In a small bowl, stir together the pork cubes, soy sauce, and gingerroot. Cover and refrigerate the mixture while preparing the remaining ingredients.

Combine the sherry, brown sugar, ketchup, Tabasco sauce, and cornstarch in a small bowl or cup, stirring until the cornstarch dissolves. Set aside.

Trim off the bottom 2 or 3 inches (coarser, stalk-like part) of the celery-cabbage; reserve for making a soup or stew. Cut the remainder (leafy portion) of the head in half lengthwise; then cut crosswise into 1-inch-wide pieces. Rinse and drain the celery-cabbage pieces thoroughly. Trim off the scallion green tops; chop and reserve for the garnish. Cut the white part of the scallions into 1-inch lengths and set aside separately.

Heat the oil in a very large (12-inch diameter or larger) skillet or sauté pan over high heat until hot but not smoking. Add the garlic and white scallion pieces and cook, stirring, for 30 seconds. Remove the pork cubes from the marinade with a slotted spoon and add to the skillet. (Reserve the marinade.) Cook, stirring, for 2 minutes. Lower the heat to medium-high and cook the pork, stirring, for 2 minutes longer. Add the celery-cabbage to the skillet, stirring. (At first the leaves will overfill the pan, but they will gradually decrease in volume.) Cook, continuing to stir, for 30 seconds longer. Briefly stir the sherry-cornstarch mixture and add it to the skillet, along with the reserved marinade. Cook, stirring, for 1½ to 2 minutes longer, or until the liquid is slightly thickened and clear. Transfer the contents of the pan to a serving dish. Sprinkle the mixture with the reserved chopped scallion tops and serve.

Makes 4 servings.

VEGETABLE AND GROUND BEEF SKILLET DINNER

(main dish)

When time is short, this makes a delicious, quick meal.

1   pound lean ground beef
1   medium-sized onion, finely chopped
2   garlic cloves, minced
1   15-ounce can tomato sauce
½   cup water
½   teaspoon dried basil leaves
¼   teaspoon dried marjoram leaves
¼   teaspoon salt
⅛   teaspoon black pepper, preferably freshly ground
1½ cups thinly sliced carrots
1½ cups thinly sliced celery-cabbage (or substitute Napa or green cabbage)
2   cups diced zucchini or yellow squash
1   cup uncooked elbow macaroni

In a very large deep skillet over medium-high heat, brown the ground beef with the onion and garlic, breaking up the meat with a spoon. Drain off and discard any excess fat. Stir in the tomato sauce, water, basil, marjoram, salt, pepper, and carrots. Bring the mixture to a boil; then cover the pan tightly. Lower the heat and simmer the mixture for 5 minutes, stirring occasionally. Add the celery-cabbage and zucchini, and continue simmering for about 5 to 7 minutes longer, or until all the vegetables are tender but not mushy.

Meanwhile, cook the macaroni according to the package directions (omitting the salt, if desired) until it is "al dente" (slightly firm). Drain the macaroni well and set it aside. When the meat-vegetable mixture is cooked through, gently stir in the macaroni. Heat, while stirring gently, about 1 minute longer; then serve.

Makes about 4 servings.

CHICKEN AND CHINESE-STYLE VEGETABLES

(main dish or side dish)

This dish is very versatile, and can be adapted to a variety of vegetables if those called for in the recipe are unavailable.

¾ cup chicken broth or bouillon (reconstituted from cubes or granules)
¼ cup dry sherry
2 tablespoons soy sauce
1 teaspoon sugar
½ teaspoon ground ginger
2 tablespoons vegetable oil
1 large yellow onion, thinly sliced
1 garlic clove, minced
1 pound boned and skinned chicken breast meat (from about 4 medium-sized breast halves), cut into ½-inch cubes
2 large bok choy stalks, including leaves, thinly sliced on the diagonal
3 celery-cabbage stalks, including leaves, thinly sliced on the diagonal
1 medium-sized scallion, including green top, thinly sliced on the diagonal
1 medium-sized sweet red pepper, thinly sliced (if unavailable, substitute sweet green pepper)
2 cups fresh snow pea pods, stem ends removed (or 1 6- to 10-ounce package frozen snow pea pods, thawed and drained)
1 8-ounce can sliced water chestnuts or bamboo shoots, drained
¼ cup cashew pieces or slivered almonds, preferably salt-free
2 tablespoons cornstarch
2 tablespoons cold water

To Serve
Hot cooked white or brown rice

In a small bowl or measuring cup, combine the chicken broth, sherry, soy sauce, sugar, and ground ginger. Set aside.

In a very large skillet or a wok, heat the oil over medium-high heat; then sauté the onion until it is tender but not browned. Add the garlic and chicken and sauté (or stir-fry) until the chicken is opaque. Add the bok choy, celery-cabbage, scallion, pepper, snow pea pods, water chestnuts, and cashews. Stir-fry for about 1 minute. Pour in the chicken broth mixture, cover the pan tightly, and steam the mixture for about 3 to 5 minutes, or just until the vegetables are crisp-tender. Do not overcook.

Mix the cornstarch with the water and add the mixture to the pan, stirring constantly. Heat just until the sauce is thick and bubbly. Serve over hot cooked rice.
Makes 4 main-dish servings.

Variation for side dish

CHINESE-STYLE MIXED VEGETABLES

For a delicious side dish, simply omit the chicken. This makes 8 side-dish servings.

CHICKEN WITH CABBAGE AND NOODLES

(main dish)

2 medium-sized leeks
1 pound boned and skinned chicken breast meat (from about 4 medium-sized breast halves), OR 1 pound raw turkey breast cutlets
5 tablespoons dry sherry or dry white wine, divided
2 tablespoons butter or margarine
1 medium-sized onion, finely chopped
1 pound green or Savoy cabbage, shredded or finely chopped
¼ teaspoon salt
⅛ teaspoon black pepper, preferably freshly ground
¼ teaspoon dried thyme leaves
2 tablespoons water
4 cups uncooked medium-wide flat noodles

Clean the leeks carefully to remove any sand inside them. (To do this, cut off and discard the root and all the green except for 1 inch. Then cut the leek in half lengthwise and separate the "layers." Rinse the layers well under cold running water; then drain the leeks on paper towels.) Chop the leeks finely and set them aside.

Cut the chicken into ¾- to 1-inch pieces and toss it with 2 tablespoons of the sherry. Set aside.

In a large skillet, over medium-high heat, melt the butter; then add the onion and reserved leeks and cook until tender but not browned. Add the cabbage and stir until it is wilted. Then mix in the chicken and its marinade, the remaining 3 tablespoons of sherry, salt, pepper, thyme, and water. Bring to a simmer; then lower the heat and cover the skillet tightly. Steam the chicken and vegetables for about 10 minutes, or until they are tender.

Meanwhile, cook the noodles according to the package directions and drain them well. When the chicken mixture is cooked through, stir in the noodles and serve.

Makes about 4 servings.

Carrots

Although carrots have been around for thousands of years, the healthful, bright orange carrot we eat today was probably not developed until the Middle Ages, and came into popular usage only in the seventeenth century. A native of Asia, the ancient ancestor of the modern carrot was likely a pale or purplish, tough, bitter taproot of an herb. Today, most of the wild carrots thriving throughout the world appear to derive from cultivated plants that reverted to type.

The carrot is a member of the parsley family, which includes not only parsley and dill, but parsnips, celery, and fennel, as well. It is also related to the wild flower known as Queen Anne's Lace. Interestingly, carrot foliage and flowers were once thought to be so attractive that stylish women in the English court used them to adorn their hair and hats.

Carrots arrived in America with the early English colonists, and were quickly accepted by the native Indians who not only incorporated them into their diet but began to grow them throughout the country.

Cultivated carrots come in an amazing variety of colors and shapes—including orange, yellow, red, white, purple, long, short, slender, stubby, pointed, blunt, and even bulbous. However, the only carrots grown widely in this country are the slender, pointed, orange ones. Large, mature carrots fitting this description are most commonly sold. "Baby" carrots (which are not fully grown) and the mature, miniature variety known as "Belgian carrots" are considered to be more of a delicacy.

Among the vegetables, carrots are second only to beets in the amount of natural sugar they contain, which is probably why they are so often baked into desserts.

Though carrot tops, or greens, are rarely eaten, they are edible both raw and cooked, and are actually a good source of nutrients.

Availability: Packaged fresh carrots, which have been trimmed of their tops, are plentiful year round just about everywhere in the United States. Loose carrots with their green tops intact are often only available in specialty markets or in farmers' market during the summer and fall.

Choosing the Best: Look for firm, smooth, bright-orange carrots, which are not flabby, soft, or shriveled. Avoid carrots with greenish areas at the top and those that are very thick. The latter are often very tough and fibrous in the center. Also, avoid carrots that have begun to split or those with yellow shoots at the top, as these are overly mature.

If choosing carrots with the tops still intact, look for greens in good condition as a sign of freshness.

Nutritional Value: Carrots are exceptionally rich in beta-carotene, a vitamin precursor which converts to vitamin A in the human body. (Beta-carotene may play a role in helping to prevent lung cancer.) Carrots also contain some vitamin C and small amounts of other vitamins and minerals. They are also a good source of non-nutritive fiber.

Storage: If the green tops are removed, and carrots are wrapped to prevent loss of moisture, they will keep quite well for several weeks, even months, in the refrigerator. (The greens absorb moisture from the carrots and, thus, hasten shriveling.) The amount of beta-carotene in carrots actually increases with storage.

Preparation and Basic Cooking: Trim the greens (if present), stem end (and any greenish part), and root tip from the carrots. Young, fresh, tender carrots need only be scrubbed well, but not peeled unless desired. When peeling carrots, remove only a very thin layer, as many nutrients are concentrated in and right under the skin.

Carrots are excellent eaten raw, whole, cut into sticks, or shredded into salads. Carrots may be cooked whole or cut into large chunks (best for slow-cooking dishes, such as stews); sliced crosswise into circles or on the diagonal into oval-shaped pieces; diced; or cut into julienne or tiny matchstick strips.

To cook carrots, put them into a saucepan with a small amount of water. Cover and bring the water to a boil over high heat. Immediately lower the heat and simmer just until the carrots are tender. Thin slices will take about 7 to 10 minutes. Do not overcook the carrots or they will get mushy.

Simple Serving Suggestions: For extra flavor, carrots may be cooked with a few teaspoons of honey added to the cooking water. Or substitute orange juice for the cooking water. Serve the carrots plain or with a touch of butter and perhaps some shredded orange peel. Cooked carrots also go well with a sprinkling of chopped herbs, such as parsley, dill, mint, or thyme.

Raw carrot sticks are perfect with all sorts of dips. Prepare a quick salad by shredding raw carrots and tossing in a few tablespoons of olive oil, lemon juice, and/or cider vinegar, and a pinch each of salt and pepper. Shredded carrots also have an affinity for plumped raisins.

VEGETABLE STOCK

This stock can be used wherever vegetable broth or bouillon is called for in this book. It can also serve as a base for homemade vegetable sauces.

1   tablespoon butter or margarine
5   carrots, coarsely sliced
3   celery stalks, coarsely sliced
3   large onions, coarsely chopped
1   large turnip, coarsely chopped
2   cups coarsely chopped parsley leaves
1½ cups coarsely chopped cabbage
8   cups water
1   large bay leaf
1½ teaspoons salt
¼   teaspoon whole black peppercorns
    Pinch of ground allspice
    Pinch of ground thyme

Melt the butter in a 5-quart or larger pot over medium-high heat. Add the carrots, celery, onions, turnip, and parsley leaves and cook, stirring, for 8 to 10 minutes, or until the vegetables are heated through but not browned. Stir in all the remaining ingredients and bring the mixture to a boil. Lower the heat and gently simmer the mixture, covered, for about 1½ hours, or until the liquid has been reduced by almost half. Strain the stock through a fine sieve. Cover and refrigerate the stock for several days or freeze it for longer storage.

Makes about 5 cups.

CREAMY CARROT SOUP

This tasty, peach-colored soup is a winner. Don't let the turnips give you pause; they add wonderful flavor to this tasty soup—and not a soul will know they're there.

2	tablespoons butter or margarine
1	medium-sized onion, finely chopped
1	pound carrots, thinly sliced (about 3 cups)
4	small turnips, peeled and diced (about 2 cups)
1	large potato, peeled and diced (about 1 cup)
2	cups chicken broth
¼	teaspoon dried thyme leaves
¼ to ½	teaspoon salt (use less with seasoned chicken broth)
⅛	teaspoon black pepper, preferably freshly ground
	Pinch of ground nutmeg
1	13-ounce can evaporated regular, lowfat, or skimmed milk
1 to 1½	cups regular, lowfat, or skim milk

Garnish (optional)
1	sweet red pepper, very thinly sliced

In a 3-quart saucepan, over medium-high heat, melt the butter; then cook the onion until it is tender but not browned. Stir in the carrots, turnips, potato, and broth. Bring to a boil; then lower the heat so that the mixture simmers. Cover tightly and cook for about 20 minutes, or until the vegetables are very tender. In a food processor or blender, or with a food mill, in batches if necessary, purée the vegetables and their cooking liquid.

Return the purée to the saucepan and stir in the thyme, salt, pepper, nutmeg, and evaporated milk. Add enough regular-strength milk for a desired consistency. Cook the soup over medium heat, stirring often, until it is heated through. If desired, top each serving with a few slices of sweet red pepper.

Makes 6 to 8 servings.

Note: Leftover soup is delicious when reheated. However, the soup tends to thicken upon standing, so it may be necessary to add more milk.

CARROT-CABBAGE SLAW

Vegetables
2 cups shredded or grated cabbage
3 medium-sized carrots, shredded or grated

Dressing
½ cup commercial buttermilk
⅓ cup mayonnaise
1 teaspoon sugar
1 teaspoon instant minced onions
¼ teaspoon celery salt
   Scant ⅛ teaspoon black pepper, preferably freshly ground
   Pinch of garlic powder

In a medium-sized bowl, combine the grated cabbage and the carrot.

In a small bowl or cup, combine all the dressing ingredients, blending thoroughly with a wire whisk.

Stir the dressing into the carrot-cabbage mixture and mix well. Cover and refrigerate the slaw for about 1 hour to allow the flavors to blend.

Makes 6 to 7 servings.

MARINATED CARROT AND SWEET PEPPER SALAD

1½   pounds carrots, cut crosswise into ⅛-inch-thick slices
1   small sweet green pepper, diced
1   small sweet red pepper, diced (optional)
3 to 4 scallions, including green tops, finely chopped
3   tablespoons chopped fresh chives (or 1½ tablespoons dried chives)

Marinade
½   cup apple cider vinegar
3½   tablespoons sugar
3   tablespoons tomato paste
2   tablespoons vegetable oil
1   tablespoon water
1½   teaspoons Worcestershire sauce
½   teaspoon powdered mustard
¼   teaspoon celery seeds
⅛   teaspoon salt
⅛   teaspoon black pepper, preferably freshly ground

Put the carrots into a medium-sized saucepan and just barely cover them with water. Cover and bring the water to a boil over high heat. Lower the heat and simmer for 5 to 8 minutes, or until the carrots are just barely tender. Remove the carrots from the heat and turn out into a colander. Rinse them under cold water and let stand until they are thoroughly drained.

Combine the drained carrots, sweet peppers, scallions, and chives in a large noncorrosive bowl. Stir together all the marinade ingredients in a small bowl or cup. Pour the marinade over the vegetables and toss until well mixed. Refrigerate the salad, covered, for at least 2 hours and up to 3 days before serving. Toss the salad several times during storage.

Makes 6 to 8 servings.

CREAMY MOLDED CARROT-PINEAPPLE SALAD

2   packets unflavored gelatin
1⅔ cups boiling water
1   8-ounce can juice-packed crushed pineapple, well drained (juice reserved)
⅓   cup sugar
2   tablespoons lemon juice, preferably fresh
⅓   cup frozen orange juice concentrate
½   cup part-skim ricotta (if unavailable, substitute regular ricotta)
3   tablespoons plain lowfat yogurt
⅛   teaspoon ground nutmeg
2   medium-sized carrots, shredded or grated
1   celery stalk, thinly sliced
½   cup finely chopped or grated cabbage

Put the gelatin, water, and reserved pineapple juice into a blender container and process on medium speed until well mixed, about 15 seconds. Add the sugar, lemon juice, orange juice concentrate, ricotta, yogurt, and nutmeg. Blend on medium-high speed until thoroughly combined and the ricotta is smooth, about 20 to 30 seconds. Pour the mixture into a large bowl. Add all the remaining ingredients. Chill until thickened and syrupy but not stiff or set. Stir to evenly distribute the vegetables and fruit. Pour into an oiled 1½-quart mold. Chill the mold until the gelatin is set, about 1½ hours.

Makes 6 to 8 servings.

DILLED CARROTS

(side dish)

An easy, but very good way to dress up fresh carrots.

1 pound carrots, cut crosswise into ⅛-inch-thick slices
½ cup chicken broth or bouillon (reconstituted from cubes or granules)
1 tablespoon finely chopped fresh dillweed (or 1½ teaspoons dried dillweed)
2 teaspoons butter or margarine
⅛ teaspoon black pepper, preferably freshly ground

Combine the carrots, chicken broth, and dillweed in a medium-sized saucepan over medium-high heat. Cover and bring to a boil. Lower the heat and simmer for 6 to 8 minutes, or until the carrots are just barely tender. With a slotted spoon, transfer the carrots from the pan to a serving dish. Raise the heat to high and boil the pan liquid until it is reduced to about 1 tablespoon. Add the butter and pepper to the reduced liquid and heat until the butter melts. Add the mixture to the carrots, tossing until they are well coated.

Makes 4 to 5 servings.

SPICY GLAZED CARROT STICKS WITH APRICOTS

(side dish)

Here's an easy dish that's sure to be a favorite. To quickly cut the carrots into sticks, first slice them crosswise into 2-inch-long pieces; then cut each piece length-wise into eighths, fourths, or halves, depending on its thickness. Each stick will be almost triangular in cross section. (For a more elegant presentation, the carrots may be cut into a finer julienne, in which case they will take less time to cook.)

1½   tablespoons butter or margarine
1   medium-sized onion, halved and very thinly sliced
1   pound carrots, cut into approximate ½- by 2-inch sticks (see note above)
3   ounces dried apricots (about 15), cut lengthwise into thirds
½   cup orange juice
¼   teaspoon ground cinnamon
¼   teaspoon ground cloves
1   tablespoon packed light or dark brown sugar

In a medium-sized saucepan over medium-high heat, melt the butter; then cook the onion, stirring, until it is tender but not browned. Add the remaining ingredients and bring to a boil, stirring. Lower the heat, cover the saucepan, and simmer the carrots, stirring occasionally to evenly distribute the sauce. Cook the carrots for about 20 minutes, or until they are tender. Remove the cover and raise the heat. Stirring constantly, rapidly boil down the sauce until it forms a rich glaze on the carrots.

Makes 4 to 6 servings.

CRAZY QUILT BROWN RICE

(side dish)

Many different flavors and textures combine to produce this tasty rice and vegetable pilaf.

2   tablespoons butter or margarine
1   medium-sized onion, finely chopped
1   cup uncooked brown rice
1   medium-sized carrot, finely chopped
1   medium-sized sweet green pepper, finely chopped
2   celery stalks, finely chopped
1   cup chopped fresh mushrooms
1   medium-sized tart apple, cored and finely chopped (peeling is optional)
1¾ cups chicken broth or chicken or vegetable bouillon (reconstituted from cubes or granules)
¼   cup slivered almonds
½   teaspoon dried thyme leaves
½   teaspoon ground (rubbed) sage
⅛   teaspoon black pepper, preferably freshly ground

In a 2½- to 3-quart skillet over medium-high heat, melt the butter; then cook the onion, stirring, until it is tender but not browned. Add the rice and cook, stirring, for 1 minute longer. Stir in the carrot, green pepper, celery, and mushrooms and cook for 2 minutes longer. Then add the remaining ingredients.

Bring to a boil and cover. Lower the heat and simmer the rice for about 40 to 45 minutes, or until all the liquid has been absorbed. Stir before serving to evenly distribute the vegetables.

Makes about 6 servings.

POT ROAST WITH ORANGE VEGETABLES

(main dish)

The vegetables in this stew are a great source of the nutrient beta-carotene, which is reputed to help prevent lung cancer.

1 to 2	tablespoons vegetable oil
1	2- to 3-pound boneless lean beef pot roast
1	large onion, thinly sliced
	About 4 to 5 cups water
3	medium-sized carrots, cut into 2-inch sections
2	medium-sized sweet potatoes, peeled and cut into large chunks
1	medium-sized butternut squash, peeled and cut into large chunks
1	medium-sized rutabaga, peeled and cut into large chunks
¼	cup raisins
1 to 2	tablespoons light or dark brown sugar (or to taste)

Thickener (optional)
Cornstarch and cold water

In a large Dutch oven or soup pot over medium-high heat, heat the oil; then brown the pot roast on all sides. Stir in the onion and cook until it is tender but not browned. Add just enough water to cover the roast. Bring to a boil; then lower the heat, cover, and simmer for 1½ hours. Add the carrots, sweet potatoes, squash, rutabaga, and raisins. Sprinkle the brown sugar on top. Continue simmering, covered, until the meat is very tender, about 1 hour longer (or more, if necessary).

Carefully transfer the meat to a serving platter. Use a slotted spoon to place the vegetables around the meat. If the sauce is thick enough, serve it as is. If the sauce is too thin, either quickly boil it down a bit, or thicken it with 1 to 2 tablespoons of cornstarch dissolved in an equal amount of cool water. (Add the cornstarch mixture while stirring, and bring the sauce to a boil.) Pour some of the sauce over the meat and vegetables, and pass the remainder in a small pitcher or sauceboat.

Makes 6 to 8 servings.

GARDEN MEAT LOAF

(main dish)

The carrots are an interesting and flavorful addition to this tasty meat loaf. Because the loaf is delicious either hot or cold, it can be served for dinner and the leftovers sliced for sandwiches the next day.

1	pound lean ground beef
1	medium-sized carrot, grated or shredded
1	celery stalk, grated or shredded
1	medium-sized onion, shredded or very finely chopped
1	garlic clove, minced
1	cup old-fashioned or quick-cooking rolled oats
½	cup ketchup
¼	cup finely chopped fresh parsley leaves
2	large eggs, slightly beaten
½	teaspoon dried basil leaves
½	teaspoon powdered mustard
¼	teaspoon dried thyme leaves
¼	teaspoon celery salt
2 to 3	drops Tabasco sauce
½	teaspoon salt
¼	teaspoon black pepper, preferably freshly ground

In a large bowl, combine all the ingredients and mix thoroughly. Press the mixture into an 8½- by 4½-inch loaf pan. Bake in a preheated 375-degree oven for 50 to 60 minutes, or until firm and nicely browned on top. Cool 10 minutes before slicing.

Makes 5 to 6 servings.

CHICKEN BARBECUE SANDWICHES

(light main dish)

Grated vegetables add flavor and nutritional value to this tangy barbecue mixture.

Barbecue
1   tablespoon vegetable oil
1   pound boneless chicken (meat from 3 large breast halves), cut into 1- by ½-inch pieces
1   medium-sized onion, finely chopped
1   garlic clove, minced
1   medium-sized carrot, grated or shredded
1   broccoli stem, grated or shredded (Reserve the flowerets for another use.)
1   15-ounce can tomato sauce
2   tablespoons apple cider vinegar
2   tablespoons sugar
1   bay leaf
¾   teaspoon powdered mustard
¼   teaspoon chili powder
⅛   teaspoon ground cloves
    Pinch of cayenne pepper (or to taste)
    Pinch of black pepper, preferably freshly ground

To Serve
    Toasted English muffins or hamburger buns

Heat the oil in a large heavy skillet over medium-high heat. Add the chicken, onion, and garlic. Cook, stirring, until the chicken is opaque. Add all the remaining barbecue ingredients. Lower the heat to medium-low, cover, and simmer for about 15 minutes, stirring frequently. Uncover the skillet, lower the heat to very low, and simmer for an additional 10 minutes, stirring frequently, so that the sauce can cook down slightly. Remove the bay leaf and serve the barbecue mixture on toasted open-faced muffins or buns.
   Makes 4 to 5 servings.

CARROT-BRAN MUFFINS

½   cup 100-percent bran cereal
⅔   cup water
1   cup enriched all-purpose or unbleached white flour
⅓   cup nonfat dry milk powder
2   teaspoons baking powder
¼   cup sugar
½   teaspoon ground cinnamon
¼   teaspoon ground cloves
¼   teaspoon salt
1   large egg, lightly beaten
3   tablespoons vegetable oil
½   cup grated carrot
½   cup dark raisins, dried currants, or chopped dates

In a small bowl or cup, combine the bran cereal and the water. Set aside.

In a medium-sized bowl, combine the flour, powdered milk, baking powder, sugar, cinnamon, cloves, and salt. Stir until well blended.

With a fork, stir the egg and vegetable oil into the bran mixture. Add to the dry ingredients. Stir with a large spoon just until mixed. Stir in the carrot and dried fruit. Spoon into 12 lightly greased or nonstick spray-coated muffin cups. Bake in a preheated 400-degree oven for about 17 to 19 minutes, or until lightly browned. If the muffins are difficult to remove, tap the edge of the muffin tin against the kitchen counter a few times to loosen them. Serve warm.

Makes 12 muffins.

PINEAPPLE-CARROT QUICK BREAD

1   cup enriched all-purpose or unbleached white flour
1   cup whole wheat flour
1   teaspoon baking soda
1   teaspoon baking powder
1   teaspoon ground cinnamon
¾   teaspoon ground allspice
¼   teaspoon ground nutmeg
⅛   teaspoon salt
3   tablespoons vegetable oil
1   large egg, slightly beaten
1   cup grated carrots
¾   cup sugar
½   cup juice-packed crushed pineapple, well drained (juice reserved)
⅔   cup reserved pineapple juice (orange juice may be added to make ⅔ cup juice if necessary)
½   cup golden or dark raisins

In a medium-sized bowl combine the flours, baking soda, baking powder, cinnamon, allspice, nutmeg, and salt. Stir to mix well. Set aside.

In a small bowl, combine all of the remaining ingredients, and stir to mix well. Add this liquid mixture to the dry ingredients, stirring with a large spoon to combine thoroughly.

Spoon the batter into a greased or nonstick spray-coated 9- by 5-inch loaf pan. Bake in a preheated 350-degree oven for 50 to 60 minutes, or until the top of the loaf is golden brown and a toothpick inserted in the center comes out clean.

Makes 8 to 10 servings.

CARROT SNACK CAKE

The carrots in this good and easy cake don't need to be grated; instead they are simply placed in a blender with liquid and puréed until smooth.

2	large carrots, cut into 1-inch lengths
1¼	cups orange juice
¾	cup dark raisins
1	cup packed light brown sugar
3	tablespoons vegetable oil
1⅔	cups enriched all-purpose or unbleached white flour
1	cup whole wheat flour
2½	teaspoons baking powder
1	teaspoon baking soda
2	teaspoons ground cinnamon
1	teaspoon ground nutmeg
1½	teaspoons ground allspice
1	teaspoon ground ginger
2	large eggs, lightly beaten

Glaze

1	cup confectioners' sugar (sifted if lumpy)
½	teaspoon vanilla extract
1 to 2	tablespoons orange juice

Combine the carrots and the 1¼ cups of orange juice in a blender container. Blend on high speed for 1½ to 2 minutes, or until the carrots are completely puréed and the mixture is smooth. Transfer the puréed carrots to a medium-sized saucepan. Bring to a boil over medium-high heat. Boil the carrot purée for 1 minute. Remove the pan from the heat and stir in the raisins, brown sugar, and oil. Set the mixture aside to cool slightly.

Combine the flours, baking powder, baking soda, cinnamon, nutmeg, allspice, and ginger in a medium-sized bowl and stir until thoroughly mixed. Add the puréed carrot mixture and then the beaten eggs to the dry ingredients; stir lightly until thoroughly blended but not overmixed.

Spoon the batter into a greased 11- by 7-inch baking pan or casserole, smoothing the mixture out to the edges with a spoon. Bake in a preheated 375-degree oven for 30 to 35 minutes, or until the cake is nicely browned and a toothpick comes out clean when inserted in the center. Transfer the pan to a rack and let stand for 5 minutes.

Meanwhile, prepare the glaze by stirring together the confectioners' sugar, vanilla, and enough orange juice to yield a smooth, fairly stiff mixture. Immediately spread the glaze over the cake with a knife or spatula. Allow the cake to stand until it is cool before serving. Cut the cake into rectangles and serve from the pan.

Makes 8 to 10 servings.

Cauliflower

A member of the cabbage family and a close relative of broccoli, cauliflower has long been enjoyed by peoples living in the coastal regions of the Mediterranean and in Asia Minor. The writings of the Roman naturalist Pliny speak of the vegetable in the second century B.C. But mention of its cultivation dates as far back as the sixth century B.C.

Cauliflower was introduced into medieval Europe by the Arabs, during their occupation of Spain. In fact, by the twelfth century, Spaniards were eating as many as three varieties of the vegetable.

In sixteenth-century England, cauliflower was called "Cyprus coleworts," probably because it was first imported from the island of Cyprus.

Although the technique was developed many thousands of years ago, cultivation of cauliflower for the dinner table is a fairly sophisticated procedure. The edible white head, known as the "curd," does not grow large and succulent by itself. When the bud forms in the center of the new plant, the outer leaves are gathered together into a loose tent over the developing curd. Sheltered in this way from the harsh rays of the sun, the heads grow snowy white and tender.

Availability: In many areas, cauliflower can be purchased all year round. However, the peak season is September through November.

Choosing the Best: Select compact, firm, creamy white heads that seem densely packed. Avoid heads that are spotted, bruised, or crumbling. Yellowing and spreading flowerets are a sign of overmaturity—as is a strong "cabbagy" odor. When present, leaves should be crisp and bright green, not wilted.

Nutritional Value: Cauliflower is a good source of vitamins C and A and potassium. It is also a fair source of iron.

Storage: Unwashed cauliflower should be wrapped well to prevent moisture from evaporating. It can be stored in the refrigerator for up to a week.

Preparation and Basic Cooking: Cut off any leaves; then wash the head under cool running water. Cauliflower can be cooked either whole or as flowerets or slices. A whole head should be boiled, facedown, for 15 to 30 minutes. Test for doneness by piercing the stem end with a fork. The flesh should be tender-crisp.

Flowerets can be simmered in a tightly covered saucepan in 1 or 2 inches of water. Cover and bring to a boil. Then lower the heat and simmer for 5 to 12 minutes, depending on the size of the flowerets and the degree of doneness desired. Slices should be simmered for 4 to 10 minutes. Steaming will take a few minutes longer. Pieces should be cooked to the tender-crisp stage.

Simple Serving Suggestions: To dress up cauliflower, serve with a simple butter, white or cheese sauce, or with melted cheese. A topping of fine bread crumbs lightly browned in butter can also be used as a garnish. Good seasonings for this vegetable include chives, dillweed, nutmeg, minced parsley, celery salt, and lemon juice.

Uncooked cauliflower is also tasty in tossed salads and as a crudité with other raw vegetables and a dip.

CAULIFLOWER VICHYSSOISE

This variation on a classic soup uses cauliflower (a member of the cabbage family which may help prevent cancer) instead of potatoes for a light and flavorful soup that is very low in calories.

1   bunch leeks (about 4 medium sized)
3   tablespoons butter or margarine
1   medium-sized onion, finely chopped
3   cups chicken broth or bouillon (reconstituted from cubes or granules)
1   medium-sized head cauliflower, trimmed and cut into flowerets
¼   teaspoon salt (or to taste)
⅛   teaspoon white or black pepper, preferably freshly ground
1   cup whole, lowfat, or skim milk
1   can (13 ounces) evaporated regular or skimmed milk
     Chopped fresh chives (or parsley), for garnish

Clean the leeks carefully to remove any sand inside them. (To do this, cut off and discard the root and all the green except for about 1 inch. Then cut the leeks in half lengthwise and separate the "layers." Rinse the layers well under cold running water; then drain the leeks on paper towels.) Chop the cleaned leeks finely. This should yield about 2 cups, but the exact amount is not critical.

Melt the butter in a large saucepan over medium-high heat and add the leeks and onion. Cook until they are tender but not browned, about 8 to 10 minutes, stirring often. If the leeks are browning too quickly, add a little water to the saucepan.

When the leeks and onion are tender, add the chicken broth, cauliflower, salt, and pepper to the saucepan. Bring the mixture to a boil; lower the heat and simmer, covered, for about 15 to 20 minutes, or until the cauliflower is very tender. Remove the saucepan from heat and purée the vegetables and broth in a blender, food processor, or food mill (in two or three batches, if necessary).

Return the purée to the saucepan and stir in the milk and evaporated milk. Stir the soup over medium heat just until it simmers. Cool, cover, and refrigerate the soup until it is chilled. If desired, season it with additional pepper. Stir the soup before serving. Serve it cold, with some chopped chives sprinkled on top.

Makes about 6 servings.

Note: Although it is traditionally served cold, this soup is also quite tasty when served hot.

CAULIFLOWER-DILL SOUP

This tasty soup requires a large head of cauliflower. Some flowerets and stem pieces are chopped, while the remainder of the flowerets are used whole.

2   cups finely chopped cauliflower flowerets and stem pieces
1   medium-sized potato, peeled and grated or shredded
1   medium-sized onion, finely chopped
1   garlic clove, minced
5   cups water
¼   teaspoon black pepper, preferably freshly ground
2   packets or cubes vegetable bouillon
1   teaspoon dried dillweed
½   teaspoon powdered mustard
½   teaspoon celery salt
¼   teaspoon salt
4   cups small cauliflower flowerets or floweret pieces
2½   cups instant nonfat dry milk powder
¾   cup cold water
¾   cup commercial sour cream

Combine the 2 cups cauliflower flowerets and stem pieces, potato, onion, and garlic in a large saucepan. Add the 5 cups of water, pepper, vegetable bouillon, dillweed, powdered mustard, celery salt, and salt. Stir to mix well. Bring to a boil over high heat. Add the 4 cups cauliflower flowerets. Lower the heat and simmer for 15 minutes, or until the cauliflower flowerets are tender.

Combine the milk powder and the ¾ cup cold water in a small bowl. Stir to make a smooth paste. Slowly stir this into the soup. Gently simmer the soup for 5 minutes longer, stirring frequently. Lower the heat so that the soup does not boil. Stir in the sour cream, using a wire whisk, if necessary, to combine it well. Cook for 2 or 3 minutes longer, but *do not boil* or the soup may curdle.

Makes 5 to 6 servings.

CAULIFLOWER-BROCCOLI SALAD

Dressing
¼ cup plain lowfat or regular yogurt
¼ cup commercial sour cream
¼ cup mayonnaise
¼ cup finely chopped fresh parsley leaves
2 teaspoons instant minced onions
1 teaspoon dried basil leaves
½ teaspoon dried dillweed
½ teaspoon dried marjoram leaves
½ teaspoon celery salt
  Scant ¼ teaspoon garlic powder
¼ teaspoon powdered mustard

Vegetables
3 cups small broccoli flowerets
4 cups small cauliflower flowerets

To Serve
  Lettuce or other salad greens

Combine all the dressing ingredients in a medium-sized bowl.

Meanwhile, combine the vegetables and 1 or 2 inches of water in a small saucepan over high heat. Cover and bring to a boil. Lower the heat and simmer the vegetables for about 5 minutes, or until they are just crisp-tender. Drain the vegetables well in a colander. Stir them into the dressing mixture. Cover and refrigerate, tossing occasionally, for several hours to allow the flavors to blend. Serve on salad greens.

Makes 6 to 8 servings.

CAULIFLOWER WITH CHEESE SAUCE

1 medium-to-large head cauliflower, trimmed and cut into small flowerets
2 tablespoons butter or margarine
2 tablespoons enriched all-purpose or unbleached white flour
⅛ teaspoon onion powder
¼ teaspoon salt
⅛ teaspoon black pepper, preferably freshly ground
1 cup lowfat milk
3 ounces sharp Cheddar cheese, grated (¾ cup packed)
  Paprika for garnish

Combine the cauliflower with 1 or 2 inches of water in a medium-sized saucepan. Cover and bring to a boil over high heat. Lower the heat and cook for about 5 to 12 minutes, or until it is crisp-tender. Drain well in a colander.

Meanwhile, melt the butter in a small saucepan over medium-low heat. Remove from the heat. Blend in the flour, onion powder, salt, and pepper. Gradually add the milk, stirring until well mixed. Return the pan to the heat and cook over medium-low heat, stirring constantly, until the sauce is thick and smooth. Turn the heat to low and cook, uncovered, for an additional 3 minutes, stirring frequently. Stir in the cheese, blend well, and stir and cook until the cheese melts. With a large spoon, transfer the cauliflower to a serving dish. Pour the cheese sauce over the top, garnish with paprika, and serve at once.

Makes 6 to 7 servings.

CURRIED CAULIFLOWER WITH PEAS

(side dish)

Though reminiscent of Indian cuisine, this is milder than many authentic curries. The spices can be adjusted to taste.

1½	tablespoons butter or margarine
1	large onion, finely chopped
2	garlic cloves, minced
1 to 1½	teaspoons curry powder
½	teaspoon ground turmeric
¼	teaspoon ground ginger
⅛	teaspoon ground cinnamon
⅛	teaspoon ground cardamom (optional)
1	medium-sized head cauliflower, trimmed and cut into flowerets
1	cup chicken broth or bouillon (reconstituted from cubes or granules)
1	cup loose-pack frozen peas, slightly thawed
	Pinch of salt (optional)

In a large skillet over medium-high heat, melt the butter. Cook the onion and garlic, stirring, until they are tender but not browned. Add the curry powder, turmeric, ginger, cinnamon, and cardamom (if used), and stir for about 1 minute longer. Stir in the cauliflower flowerets so that they are all coated with the onion mixture. Add the broth to the skillet and bring it to a boil. Cover the skillet, lower the heat, and simmer the cauliflower, basting it often with the pan juices, for about 8 to 15 minutes, or until it is just crisp-tender. Stir the peas in with the cauliflower and continue simmering, covered, for about 2 to 4 minutes longer, or just until the peas are heated through. If desired, season with the salt.

Makes about 6 servings.

STEAMED CAULIFLOWER WITH PIMIENTO SAUCE

(side dish)

Unusual and eye-catching, this dish features creamy white cauliflower napped in an orange-gold sauce. The flavor of the pimiento and tomato in the sauce complements the cauliflower nicely.

1   large head cauliflower, trimmed and separated into large flowerets (about 6 to 7 cups)
2   tablespoons butter or margarine
2   tablespoons chopped red onion
2   tablespoons all-purpose enriched or unbleached white flour
⅓   cup well-drained chopped canned pimiento
1   large tomato, peeled, seeded, and coarsely chopped
¾   cup whole or lowfat milk
¼   cup light cream
½   teaspoon salt
⅛   teaspoon white pepper, preferably freshly ground
    Scant ⅛ teaspoon cayenne pepper
1   tablespoon finely chopped fresh chives for garnish (optional)

Put about 1 inch of water into a large saucepan or steamer fitted with a steamer basket and bring to a boil over medium-high heat. Add the cauliflower flowerets to the basket. Cover the pan, lower the heat slightly, and steam the cauliflower for 7 to 10 minutes, or until it is almost tender.

Meanwhile, melt the butter in a medium-sized saucepan over medium-high heat. Add the onion and cook, stirring occasionally, for 5 or 6 minutes, or until the onion is tender but not browned. Add the flour and continue cooking, stirring, for 2 minutes longer. Transfer the cooked onion mixture to a blender. Add the pimiento, tomato, and milk to the blender. Blend on medium speed for 1 to 1½ minutes, or until the mixture is completely smooth. Transfer the mixture to a clean saucepan. Add all the remaining ingredients, *except* the chives, and bring to a boil, stirring, over medium-high heat. Boil the sauce for 1 to 1½ minutes, or until it is thickened and smooth. Then remove the sauce from the heat.

Arrange the cauliflower in a serving bowl and top with the pimiento sauce. Then top the sauce with a sprinkling of chopped chives, if desired.

Makes 5 to 7 servings.

VEGETABLES AND CHEESE

(side dish)

This recipe, which can be made in a pot on top of the stove or in the microwave oven, features an easy cheese topping made from lowfat Longhorn cheese. Because this cheese has a high moisture content, it melts readily to a sauce-like consistency.

2 large potatoes, peeled and cut into 1-inch cubes
2 medium-sized onions, quartered
2 medium-sized carrots, thinly sliced
1 sweet green or red pepper, cut into 1-inch squares
3 cups small cauliflower flowerets
6 ounces lowfat Longhorn cheese, cut into small cubes

CONVENTIONAL METHOD
Combine the potatoes, onions, and carrots in a large saucepan. Cover them with water and bring to a boil over high heat. Lower the heat and simmer for 8 minutes. Add the remaining vegetables. Bring to a boil again. Lower the heat and simmer an additional 5 minutes, or until the cauliflower is just crisp-tender. Drain the vegetables well in a colander. Return the vegetables to the cooking pot. Stir in the cheese. Cover and cook over low heat for about 2 minutes, or until the cheese has melted. Stir to coat the vegetables with the cheese. Serve at once.

MICROWAVE METHOD
Combine the potatoes, onions, and carrots in a large casserole. Add about ½ inch of water to the bottom. Cover and microwave on full power for 9 to 11 minutes, turning the casserole a quarter turn once during the cooking period. Add the remaining vegetables and microwave on full power an additional 5 to 7 minutes, or until the potatoes and carrots are tender and the cauliflower is just crisp-tender. Drain the vegetables well in a colander. Return them to the casserole. Stir in the cheese. Microwave on full power an additional minute, or until the cheese has melted. Stir to coat the vegetables with the cheese. Serve at once.

Makes 6 to 8 servings.

CAULIFLOWER AND CHEDDAR CASSEROLE

(side dish or light main dish)

1   medium-sized head cauliflower, trimmed and broken (or cut) into small flowerets (5 to 6 cups)
3   ounces sharp Cheddar cheese, grated (¾ cup packed)
3   large eggs, lightly beaten
1   cup skim, lowfat, or whole milk
3   tablespoons finely chopped fresh parsley leaves
1   tablespoon instant minced onions
¼   teaspoon salt (or to taste)
⅛   teaspoon black pepper, preferably freshly ground
⅛   teaspoon powdered mustard

Put the cauliflower flowerets in a medium-sized saucepan with about 1 inch of water (or in a steamer basket over water), and bring the water to a boil over high heat. Cover the pan tightly, lower the heat, and steam the cauliflower for about 5 to 8 minutes, or just until it is barely crisp-tender. Drain the cauliflower well and transfer it to a greased or nonstick spray-coated shallow 2-quart casserole dish. Sprinkle the cheese evenly over the cauliflower.

Stir together the remaining ingredients until well combined and pour the mixture evenly over and around the cauliflower. Bake the casserole, uncovered, in a pre-heated 325-degree oven for about 30 minutes, or until the cheese has melted and the custard is set.

Makes about 6 side-dish servings or about 3 light main-dish servings.

CHICKEN AND PARSLEY DUMPLINGS

(main dish)

The interesting combination of vegetables gives this hearty one-dish meal a rich flavor.

Chicken and Vegetables
1½ tablespoons butter or margarine
1 large onion, coarsely chopped
1 large celery stalk, including leaves, coarsely chopped
1½ cups coarsely chopped cauliflower flowerets
1 small turnip, peeled and diced
¼ cup finely chopped fresh parsley leaves
½ cup sliced fresh mushrooms
1 broiler/fryer chicken (about 2½ pounds), cut into large pieces
2½ cups water
1 chicken bouillon cube
2 tablespoons chopped fresh chives (or 1 tablespoon dried chopped chives)
½ teaspoon celery seeds
¼ teaspoon dried thyme leaves
¼ teaspoon dried dillweed
1 teaspoon salt
⅛ teaspoon black pepper, preferably freshly ground
2 medium-sized carrots, diced
¾ cup loose-pack frozen green peas
2 tablespoons cornstarch
¼ cup cold water

Dumplings
1 cup enriched all-purpose or unbleached white flour
1 teaspoon baking powder
¼ teaspoon salt
3 tablespoons finely chopped fresh parsley leaves
3 tablespoons instant nonfat dry milk powder
2½ tablespoons butter or margarine, melted
½ cup water (approximately)

Melt the butter in a 4-quart (or similar) Dutch oven over medium-high heat. Add the onion, celery, cauliflower, turnip, parsley, and mushrooms. Cook, stirring constantly, until the onion is limp. Add the chicken pieces, 2½ cups water, bouillon cube, chives, celery seeds, thyme, dillweed, salt, and pepper to the pot. Cover the mixture and bring to a boil. Lower the heat and simmer the chicken for 50 to 55 minutes, or until it is just tender. Remove the chicken from the pot and set it aside until cool enough to handle.

Using a large spoon, skim off and discard any fat floating on the surface of the broth. Add the carrots and peas to the pot. Cover and simmer for 15 minutes. Stir together the cornstarch and water and set aside.

Meanwhile, prepare the dumplings. Combine the flour, baking powder, salt, parsley, and nonfat dry milk in a bowl. Stir in the butter and enough water to form a soft and moist, but not runny, dough, but do not overmix. Set aside.

Remove the cooled chicken meat from the bones and cut it into bite-sized pieces. Return the meat to the pot. Stir the cornstarch mixture briefly; then add it to the pot. Allow the mixture to return to a simmer. When the liquid is clear and slightly thickened, top it with the dumplings; drop them by teaspoonfuls, spacing about ½ inch apart to allow for expansion during steaming. Cover the pot and very gently simmer about 20 minutes longer, or until the dumplings are fluffy and cooked through.

Makes about 5 servings.

Celery

Although modern cooks take celery somewhat for granted, and use it primarily with other vegetables in soups, stews, and salads, it was once highly prized. In fact, celery was one of the wedding gifts received by an Empress of China during the T'ang dynasty.

The ancient Greeks thought so much of the green leafy stalks that they gave out bunches to victorious athletes. In addition, the Greeks, as well as the ancient Romans and Egyptians, cultivated celery as a medicinal herb.

Celery traces its roots to the same family as the parsnip and the carrot. Wild celery, like wild carrot, is bitter and inedible. However, once cultivated, the wild variety yielded 2 separate plants—celery and celeriac or celery root.

Today, the most common type of celery grown in the United States is Pascal—a tall, crisp, dark green variety. Several whiter types, such as Golden Heart and White Ice (grown under paper to prevent chlorophyll from forming and turning the celery green), are not widely available because of shorter shelf life. However, they are sometimes sold as packaged "celery hearts."

Celeriac, celery knob, or celery root is a variety cultivated for its enlarged root—not for the stalks and leaves. The root, which must be peeled first, is eaten raw in salads or cooked by itself or in combination with other vegetables in soups and stews.

Availability: Celery can be purchased the year round—with the supply remaining fairly constant throughout the year.

Choosing the Best: Select crisp, firm stalks with dark leaves. Avoid bunches with limp, cracked, bruised, or loosened stalks or with brown leaves. Bunches with small dark green stalks may have a slightly bitter flavor.

Nutritional Value: Although celery is high in potassium, its chief claim to nutritional fame is its low calorie count. Three and a half ounces of raw celery have just 17 calories. Celery is also high in fiber.

Storage: Celery loses moisture easily and so should be well wrapped before refrigerating. It can be kept for up to two weeks.

Basic Preparation: Wash celery well under running water to remove any dirt that may have collected at the base of the stems.

Simple Serving Suggestions: Crisp and crunchy raw celery is excellent in all sorts of salads—from tossed salad to tuna and egg salad. It is also popular served with other raw vegetables and a dip or stuffed with cream cheese or cottage cheese and herb mixtures.

But don't overlook the subtle flavor of cooked celery. Many experts recommend braising as the preferred method of cooking. To braise a pound of celery (about 12 to 14 stalks), trim off the tops and leaves as well as ½ inch from the bottom of each stalk. Cut the celery in half lengthwise, then into 3- to 4-inch lengths. Place in a saucepan along with ½ cup chicken bouillon (reconstituted from cubes or granules) or stock, ¼ teaspoon salt, and 1 tablespoon of butter. Cover tightly and simmer for 25 minutes, stirring occasionally, or until the celery is tender. The remaining liquid may be boiled down and served with the celery. This makes about 4 servings.

Cooked celery can be seasoned with lemon juice, butter, thyme, chopped parsley, or tarragon.

STUFFED CELERY

(appetizer)

While this recipe calls for regular cream cheese, it can also be made with the slimmed down "lowfat" cream cheese now available in many supermarkets. However, if the lowfat type is used, stir very carefully, because it tends to become "soupy" if mixed too vigorously.

Cheese Mixture
- 1   8-ounce package cream cheese at room temperature
- ½   cup juice-packed crushed pineapple, well drained
- ¼   cup grated or shredded carrot
- 2 to 3   tablespoons finely chopped sweet green pepper
- 2   tablespoons finely chopped fresh parsley leaves
- ½   teaspoon instant minced onions
- 2 to 3   drops Tabasco sauce

To Serve
- 1   bunch celery stalks, trimmed and cut into 3-inch lengths

In a medium-sized bowl, combine all the ingredients, *except* the celery. Mix well, using a wooden spoon. Make sure that no large lumps of cheese remain. With a teaspoon, spread the mixture into the celery sticks, packing well. Refrigerate until serving time.

Makes enough to stuff 1 bunch of celery.

CREAMY CELERY SOUP

     3   tablespoons butter or margarine
     1   cup vegetable or chicken bouillon (reconstituted from cubes or
         granules) or vegetable stock (page 86)
    ½   cup dry white wine
     4   cups thinly sliced celery, including some leaves
     1   large onion, finely chopped
    ½   teaspoon dried thyme leaves
    ½   teaspoon dried marjoram leaves
    ¾   teaspoon salt
    ¼   teaspoon celery salt
    ¼   teaspoon black pepper, preferably freshly ground
2 to 3   drops Tabasco sauce
     4   cups whole milk, divided
     2   tablespoons cornstarch

In a large saucepan, combine the butter, bouillon, white wine, celery, onion,
thyme, marjoram, salt, celery salt, pepper, and Tabasco sauce. Bring to a boil
over medium-high heat. Lower the heat, cover, and simmer, stirring occasionally, for about 20 to 25 minutes, or until the vegetables are tender. In batches,
if necessary, purée the vegetables in a food processor or blender until no large
pieces remain. If a food processor is used, the texture may not be perfectly
smooth.

Return the mixture to the pan over medium-high heat. Stir 3½ cups of the milk
into the celery mixture. In a small bowl or cup, combine the remaining ½ cup of the
milk with the cornstarch. Stir this into the soup. Continue stirring until the soup
boils and thickens. Lower the heat and continue to simmer, stirring frequently, for
an additional 5 minutes.

Makes 6 to 8 servings.

SAVORY CELERY AND TOMATOES

(side dish)

     2   tablespoons butter or margarine
    2½ cups ¼-inch-thick celery slices
     1   medium-sized onion, finely chopped
     1   16-ounce can tomatoes, including juice
     1   teaspoon sugar
     1   teaspoon dried basil leaves

¼ teaspoon salt
⅛ teaspoon black pepper, preferably freshly ground

In a medium-sized saucepan, combine the butter, celery, and onion over medium-high heat. Cook, stirring, until the onion is soft. Add the juice from the tomatoes (reserving the tomatoes), along with the sugar, basil, salt, and pepper. Cover, lower the heat, and simmer for about 20 minutes, or until the celery is tender. Add the tomatoes, breaking them up with a spoon. Raise the heat slightly and simmer, uncovered, for an additional 5 minutes, or until the flavors are blended.
Makes 4 to 5 servings.

QUICK SIDE-DISH STUFFING

(side dish)

This savory stuffing is designed for microwave cooking.

1 large onion, coarsely chopped
4 tablespoons butter or margarine, cut into ½-inch chunks
2 celery stalks, including leaves, thinly sliced
1 medium-sized carrot, grated or shredded
½ cup finely chopped fresh parsley leaves
½ cup chicken or vegetable bouillon (reconstituted from cubes or granules) or vegetable stock (page 86)
¼ teaspoon dried sage leaves
½ teaspoon dried thyme leaves
1 teaspoon dried marjoram leaves
¼ teaspoon black pepper, preferably freshly ground
½ teaspoon celery salt
½ teaspoon powdered mustard
8 cups bread cubes, cut from fresh or slightly stale whole wheat or white bread

In a 2½-quart casserole, mix together the onion, butter, celery, carrot, parsley, bouillon, and seasonings. Cover and place in a microwave oven. Cook for 5 to 7 minutes on full power, turning the casserole one quarter turn and stirring the vegetables once during the cooking period. Remove the casserole from the microwave. The vegetables should be almost tender. Add the bread cubes and stir very well to coat them thoroughly with the butter, bouillon, and vegetable juices.

Cover the casserole and return it to the microwave. Cook an additional 2 to 3 minutes on high power, until the stuffing is cooked through.
Makes 5 to 6 servings.

SPICY FILLING FOR PITA BREAD

(light main dish)

The unusual combination of spices gives this filling a tangy flavor. It is served with Middle Eastern pocket or pita bread rounds, which are cut in half so the meat mixture can be spooned inside. The filled pita pockets are then eaten like sandwiches.

Filling

1	pound lean ground beef
1	medium-sized onion, finely chopped
1	large garlic clove, minced
3	celery stalks, thinly sliced
1	medium-sized carrot, grated or shredded
½	cup dark raisins
1	8-ounce can tomato sauce
1	bay leaf
½	teaspoon dried thyme leaves
½	teaspoon dried marjoram leaves
½	teaspoon ground cinnamon
¼	teaspoon ground cloves
¼	teaspoon ground ginger
¼	cup chopped fresh parsley leaves
½	teaspoon salt
¼	teaspoon black pepper, preferably freshly ground
2 to 3	drops Tabasco sauce

To Serve

Medium-sized or large whole wheat or white pita bread rounds, cut in half to form semicircles

In a large skillet, cook the ground beef, onion, garlic, and celery over medium-high heat, breaking up the meat with a spoon, until the meat is brown and the onion is tender. Drain off and discard any excess fat. Stir in the carrot and raisins. Add all the remaining ingredients, stirring to combine well. Bring to a boil. Turn the heat to low, cover, and simmer for about 20 minutes, stirring occasionally, or until the flavors are well blended. Remove the bay leaf. Serve in pita bread pockets.

Makes 6 to 8 servings.

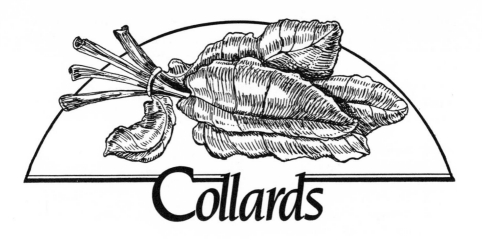

Collards

Like its close relative, kale, this primitive, leafy member of the cabbage family has changed little since it was first eaten by prehistoric man. Also like kale, the collard plant was enjoyed by the Greeks and later the Romans, whose invading armies may have introduced it into Britain and France. Collards remained the primary winter vegetable crop in England for centuries and seeds were brought along with settlers to America in the 1600s.

Extremely hardy and adaptable to both hot and cold climates, collards are unfussy growers and abundant producers of "greens," or leaves. These develop as open rosettes of large, bluish-green, cabbage-like foliage at the ends of 2- to 4-foot stalks. Collard greens win the nutritional prize for vegetables in the cabbage family, containing not only a wealth of vitamin C but twice as much vitamin A as broccoli and nearly eight times more than cabbage!

Despite their long history and nutritional benefits, collards have never gained wide acceptance in the United States. However, they are a popular winter vegetable crop throughout the South and have a prominent place in "soul food" cookery. Tall, husky stalks of "Georgia collards," one common variety, can be spotted during fall and winter in garden after garden in rural Southern black communities.

Availability: Fresh collards can be difficult to find in many parts of the North, but are generally available year round elsewhere. The peak period is January through April. Collard greens are also sold frozen and canned in some supermarkets. Though not as succulent as fresh greens, frozen ones are acceptable; however, canned collards are not likely to appeal to most tastes.

Choosing the Best: Look for bright, blue-green leaves that are unblemished and crisp. The smaller leaves are more tender and milder in flavor. Greens that are wilted and yellow have lost both flavor and nutrients.

Nutritional Value: Collard greens are among the most healthful vegetables. They are packed with vitamins A and C and also contain large amounts of calcium and phosphorus.

Storage: Collards should be loosely packed in a plastic bag and stored in the crisper. They will keep for 3 or 4 days.

Preparation and Basic Cooking: Strip or cut the leafy parts from the tough stems and midribs. Then rinse and drain the leaves. Simmer them in a small amount of water for about 15 minutes, or until they are just tender. Southern cooks often add a piece of salt pork or ham hocks to the cooking water for flavor. (Traditionally, cornbread is then served along with the pork, greens, and pot liquor.) Collard greens can also be steamed in a steamer basket for about 20 to 25 minutes.

Simple Serving Suggestions: Season collards with a little butter, salt, and pepper and perhaps a sprinkling of Parmesan cheese.

VEGETABLE-RICE SOUP

2   tablespoons butter or margarine
1   large onion, finely chopped
1   medium-sized garlic clove, minced
3   cups vegetable bouillon (reconstituted from cubes or granules) or vegetable stock (page 86)
3   cups water
1   celery stalk, thinly sliced
1   large carrot, thinly sliced
½   cup *uncooked* white rice
¼   teaspoon black pepper, preferably freshly ground
¼   teaspoon powdered mustard
¼   teaspoon dried thyme leaves
2   bay leaves
½   teaspoon dried basil leaves

¼ teaspoon dried oregano leaves
½ teaspoon dried marjoram leaves
¼ teaspoon celery salt
¼ teaspoon salt
4 cups coarsely chopped collard greens
1 cup frozen loose-pack corn kernels

In a large heavy saucepan or small Dutch oven, melt the butter over medium-high heat. Add the onion and garlic and cook, stirring, until the onion is tender. Add the bouillon and water along with the celery, carrot, and rice. Then add all the remaining ingredients, *except* the collard greens and the corn. Bring to a boil. Cover, lower the heat, and simmer for about 20 minutes, or until the rice is tender. Add the collard greens and the corn. Stir to mix well. Allow the mixture to return to a boil. Cover, lower the heat, and simmer for an additional 15 minutes, or until the collard greens are tender.

Makes 5 to 6 servings.

COLLARDS ITALIENNE

(side dish)

An easy and savory way to prepare these healthful greens.

1½ pounds fresh collards, washed and well drained
2 tablespoons olive or vegetable oil
1 small garlic clove, minced
1 small onion, finely chopped
   Pinch of crushed hot red pepper
1 large tomato, peeled, seeded, and chopped
½ teaspoon salt
⅛ teaspoon black pepper, preferably freshly ground

Trim or tear off the stems and tough midribs from the collard leaves and discard them; cut or tear the leaves into bite-sized pieces.

Combine the oil, garlic, onion, and red pepper flakes in a large pot over medium-high heat. Cook, stirring, for 4 to 5 minutes, or until the onion and garlic are limp but not browned. Gradually stir in the collard greens until they are coated with oil; at first they will overfill the pot but will gradually decrease in volume as they cook. Stir in the chopped tomato, salt, and pepper, and let the mixture come to a boil. Lower the heat, cover, and simmer the vegetables for 12 to 13 minutes, or until the collards are almost tender. Uncover the pot and continue simmering the collards until they are tender and most of the liquid has evaporated.

Makes 4 to 5 servings.

BRAISED COLLARDS WITH HERBS

(side dish)

1½ tablespoons butter or margarine
 1 teaspoon light mustard seeds
 1 small garlic clove, minced
 ½ cup beef broth or bouillon (reconstituted from cubes or granules)
 1 teaspoon dried chopped chives
 ⅛ teaspoon dried tarragon leaves
 ⅛ teaspoon black pepper, preferably freshly ground
1¼ pounds fresh collards, stems and midribs removed, torn into bite-sized
    pieces
    Pinch of salt (or to taste)

Melt the butter in a 5- to 6-quart pot over medium-high heat. Add the mustard
seeds and garlic and cook, stirring, for about 1 minute, or until the seeds start to
pop. Stir in the beef broth, chives, tarragon, and pepper and bring the mixture to a
boil. Gradually stir in the collards; at first they will overfill the pot but will decrease
in volume as they are heated. Lower the heat and cook, uncovered, stirring occasionally, for 14 to 18 minutes, or until the collards are just tender and most of the
broth has evaporated. Add salt and serve.

Makes about 4 servings.

Corn

Although we tend to think of corn as a vegetable, it is actually a grain—in fact, the only native American grain.

Early settlers were at first reluctant to try this strange new food that the Indians grew. But it undoubtedly saved the lives of both the Pilgrims and the Jamestown colonists. In danger of starving before their own crops could be harvested, they obtained corn from the local tribesmen, and later learned how to plant and cultivate it.

Since European wheat was not easily adaptable to growing conditions in either New England or the South, corn remained the chief grain crop in the colonies.

Actually, corn or "maize" cultivation is probably as old as native American culture itself. For the Aztecs, Mayans, Incas, and various North American tribes, corn was not only a dietary staple, but also an important part of their mythology and religion. Most tribes had elaborate corn dances and ceremonies in which they asked for—or gave thanks for—good crops. In Central and South America, particularly, these rites sometimes involved human sacrifices!

The Indians used corn in a wide variety of dishes from succotash to fritters and tortillas. Some even enjoyed popcorn. Many tribes fermented the grain into a corn liquor. The Incas and the Aztecs both developed corn sweeteners.

Corn continues to be a popular American vegetable and an important agricultural crop. In fact, there are over 200 varieties of sweet corn grown in the United

States today. The majority are yellow, although white corn, such as "Silver Queen," is popular in some areas.

Availability: Fresh "corn-on-the-cob" is sold all year round in some areas, although the peak season is May through September.

Both canned and frozen corn are available year round. Frozen kernels are the better substitute for fresh corn, because they maintain natural flavor and texture better than the canned product.

Choosing the Best: Because corn loses sweetness and flavor so rapidly after picking, it's best to buy from local farm stands rather than from the supermarket. Look for ears with fresh, snug, bright green husks, moist stems, and dark brown silk. The kernels should be firm, plump, and juicy looking. There should be no spaces between the rows of kernels, and rows should be filled out to the tip. Medium-sized kernels are more tender than larger ones. Tiny kernels are a sign of immaturity. Avoid soft, limp looking ears with spots, decay, or worm damage. Ripe corn kernels give a spurt of thin, milky juice when pressed with the thumb. A thicker liquid indicates overmaturity.

Nutritional Value: Corn is a good source of complex carbohydrates. Yellow corn is high in vitamin A, while white corn has practically none.

Storage: The sooner fresh corn is eaten, the sweeter and more tender it will be. This is because the sugar in the kernels begins to change into starch as soon as the ear is detached from the plant. Corn can lose up to 50 percent of its sugar during the first 24 hours after picking. If you must hold corn, try to leave it in the husk and store in the coldest part of the refrigerator. Husked corn should be placed in perforated plastic bags before refrigerating.

Preparation and Basic Cooking: Husking fresh corn can be a messy job. You might want to do it outside—or work directly over a large grocery bag. After husking, snap off the stem end. A small vegetable brush can make removing the silk easier. Rinse the husked ears in cold water. Cook, uncovered, in rapidly boiling unsalted water for 3 to 7 minutes. (Salt toughens the kernels.) A bit of sugar can be added to the water to give the corn extra sweetness. The cooked ears should still be slightly crisp.

To roast fresh corn, pull back the outer husk to the base of the ear and remove the silk. After smoothing the husk back into place, secure it with string. Soak the ears of corn in cold, salted water for 5 minutes. Roast them in a preheated 350-degree oven for 30 minutes, or until the kernels are tender.

To cut corn kernels from the cob, start at the top of the ear and run a sharp knife straight down to the stem, using a steady downward motion. Be careful not to cut into the tough cob fibers.

Simple Serving Suggestions: Fresh corn on the cob needs nothing more to dress it up than a bit of butter, salt, and pepper. However, you might like to try a seasoned butter. Simply soften ½ cup of butter slightly and mix in any of the following: 1 teaspoon curry powder, ½ teaspoon chili powder, 1 tablespoon chopped fresh parsley leaves, or 1 tablespoon chopped fresh chives.

FIESTA CORN SOUP

Zippy and full flavored, this colorful soup features not only corn, but green chilies, tomato, and Cheddar cheese. It is quite substantial and satisfying.

2   tablespoons butter or margarine
1   small garlic clove, minced
1   small onion, chopped
3   tablespoons enriched all-purpose or unbleached white flour
4   cups chicken broth or bouillon (reconstituted from cubes or granules)
¼   teaspoon chili powder
1   pound loose-pack frozen yellow corn kernels
6   ounces sharp Cheddar cheese, coarsely shredded or grated (1½ cups packed)
1   4-ounce can chopped green chilies, well drained
1   medium-sized tomato, peeled, seeded, and diced
⅛   teaspoon white pepper (or to taste)
2   teaspoons chopped fresh chives (or 1 teaspoon dried chives) for optional garnish

Melt the butter in a 4- to 5-quart saucepan or pot over medium-high heat. Add the garlic and onion and cook, stirring, for 5 to 6 minutes, or until the onion is tender but not browned. Add the flour and cook, stirring, until well blended and smooth. Slowly stir in the chicken broth and then the chili powder until well blended. Stir in the corn. Bring the mixture to a boil over high heat. Lower the heat and cover the pan. Simmer the mixture, stirring occasionally, for 10 minutes. Scoop about 2 cups of corn and a little liquid from the pan and transfer to a blender. Blend the mixture on medium speed for 1 to 1½ minutes, or until completely puréed. Return the purée to the saucepan over medium heat. Add the cheese, chilies, and tomato to the saucepan. Heat, stirring frequently to prevent sticking, for 7 to 10 minutes, or until the cheese has melted and the soup is piping hot. Stir in the white pepper to taste. Garnish the soup pot or individual servings with chives, if desired, and serve.
   Makes 5 to 6 servings.

CORN AND RICE COMBO

(side dish)

This is very simple to make, but quite attractive and tasty. Although it is not usually necessary to wash rice before using, in this case the step is important, lending a very pleasing texture to the finished dish.

1	cup *uncooked* white rice, washed and well drained
2½	tablespoons peanut oil
1	small garlic clove, minced
1½	cups chicken broth or bouillon (reconstituted from cubes or granules)
⅛	teaspoon salt
⅛	teaspoon black pepper, preferably freshly ground
1½	cups loose-pack frozen yellow corn kernels
2½	tablespoons chopped fresh chives (or 1 tablespoon dried chives)
1	small tomato, peeled and finely chopped

Combine the rice, peanut oil, and garlic in a medium-sized saucepan over medium-high heat. Cook, stirring constantly to prevent the rice from sticking, for 4 to 5 minutes, or until it has absorbed all the oil and looks slightly crisp. Stir in the broth, salt, and pepper and bring the mixture to a boil. Lower the heat, cover, and gently simmer the mixture for 15 minutes. Stir in the corn and chives and simmer for 4 to 5 minutes longer, or until the corn is cooked through. Stir in the tomato until well mixed and just heated but not cooked.

Makes 5 to 6 servings.

MEXICAN CORN AND BEANS

(side dish or light main dish)

2	teaspoons vegetable oil
1	medium-sized onion, finely chopped
1	large sweet green pepper, chopped
2	15- to 16-ounce cans red kidney beans, well drained
2	cups loose-pack frozen corn kernels
1	16-ounce can tomatoes, including juice
1	15-ounce can tomato sauce
1 to 2	teaspoons chili powder (or to taste)
⅛	teaspoon black pepper, preferably freshly ground

In a large saucepan, heat the vegetable oil over medium-high heat. Add the onion and sweet pepper and cook, stirring constantly, until the onion is soft. Add all the remaining ingredients. Bring to a boil. Cover, lower the heat, and simmer the vegetables for about 25 minutes, stirring occasionally.

Makes about 6 servings.

MICROWAVE "CORNY" MUFFINS

These muffins can be on the table in less than 10 minutes after you think of making them; a real boon to a busy cook who wants to add a special touch to a meal. The recipe has been specially adapted for use with a microwave oven. The tops of the muffins do not brown, so they may be sprinkled with a bit of paprika for color, if desired. As with most cornbreads, these taste best shortly after they are baked.

1½   cups yellow or white cornmeal, preferably stone ground
 ½   cup enriched all-purpose or unbleached white flour (or whole wheat flour)
2½   tablespoons sugar
 1   teaspoon baking powder
 ½   teaspoon baking soda
 ½   teaspoon salt
     About ½ teaspoon chili powder (optional, or to taste)
 3   tablespoons vegetable oil
 1   large egg
 1   cup commercial buttermilk
 ½   cup canned corn kernels or thawed frozen corn kernels, drained
     Paprika (optional)

Put a paper cupcake liner into each cup of a special, microwave-proof, muffin "ring" or pan. Or, if a microwave muffin pan is not available, line six small custard cups with paper liners and arrange the cups in a circle on a microwave-proof platter or tray.

In a medium-sized bowl, stir together the cornmeal, flour, sugar, baking powder, baking soda, salt, and chili powder, if desired. Make a well in the center of the dry ingredients, and add the oil, egg, and buttermilk. Mix by hand just until the batter is completely moistened. Add the corn kernels and stir just until they are evenly mixed in.

Spoon half of the batter into the paper liners so that each is just over half full. (The muffins are made in two batches of six each.) If desired, lightly sprinkle the top of each muffin with a pinch of paprika.

Bake the muffins for approximately 2½ minutes on full power, rotating the muffin pan or platter a quarter turn about every 40 seconds during the cooking period. (If you have an automatic turntable, use it instead of rotating the cups.) The muffins are done when they have risen to the tops of the paper cups and are almost completely dry on top. (A few small, moist spots may remain.) Immediately remove the muffins from the pan or the custard cups and cool them slightly on a wire rack before serving. (This helps prevent soggy bottoms due to escaping steam.) Repeat the baking procedure for the second batch of muffins.

Makes about 12 muffins.

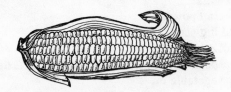

Cucumbers

The cucumber has been cultivated for more than 3,000 years. A member of the squash family, it originated in northwest India. Although served as a vegetable, it is botanically a fruit.

In biblical times, eating cucumbers was thought to offer protection against deadly insects and snakes. The ancient Greeks considered cucumbers a delicacy to be reserved for special occasions, such as victory celebrations. On the other hand, the Roman emperor Tiberius wanted them on the table every day. When he traveled, they were grown for him in a special movable frame so that his supply would not be interrupted. The French emperor Charlemagne was also fond of cucumbers, but ate them only for dessert in various pastries.

There are a number of exotic cucumber varieties. One type popular in Russia is short and thick with a tough netted brown skin. In France, white cucumbers are grown for use in soaps and other cosmetics. Japanese cucumbers are either smooth and rounded or pointed with soft spines. And there are even yellow cucumbers, which are round and about the size of an orange.

Three types are commonly available in the U.S.:

Field-grown slicing, standard, or table cucumbers grow to about 8 inches in length. Dark green in color, they are almost always studded with small white spines. (The spines come off after harvesting when the cucumbers are rubbed and washed.)

Most commercially sold standard cucumbers are sprayed with wax so that they will retain moisture and remain fresh longer.

The pickling varieties, also field grown, are much smaller than the slicing types and not quite as green in color. They have either white or black spines.

European cucumbers are smooth-skinned and generally about a foot in length. In contrast to the other varieties, they are always cultivated in greenhouses. They are generally sweeter and more tender than their field grown cousins, and are sometimes referred to as "burpless." In England, gardeners often compete to see who can produce the longest greenhouse cucumbers. Their efforts have been known to yield specimens 2 feet in length.

In the past, cucumber varieties were often bitter. But this is much less of a problem with the new hybrids.

Availability: The peak season for standard cucumbers is May through August, although they are available year round. In some areas, unwaxed cucumbers are sold during the height of the summer season.

Choosing the Best: Look for cucumbers that are slender for their length and bright green in color with whitish (as opposed to yellowing) tips. Avoid dull looking, over-large, puffy, yellowing cucumbers, as they are likely to be overmature. The flesh should be crisp and firm, not pulpy. Avoid cucumbers with soft or discolored spots.

Nutritional Value: Cucumbers have only small amounts of nutrients, but are also extremely low in calories. A whole cucumber has only about 25 calories.

Storage: Waxed cucumbers can be kept in the vegetable crisper of the refrigerator for up to a week. Unwaxed ones can be held for only a day or two.

Preparation: Waxed cucumbers should be peeled with a vegetable peeler before using. If you don't want to peel the cucumbers, you can wash off the coating, using mild soap, lukewarm water, and a vegetable brush. Unwaxed ones can be served with the peel on. Cucumbers can be sliced, diced, or cut into strips for salads. If desired, they can first be seeded. To do this, cut the cucumber in half lengthwise and scoop out the seeds with a spoon.

To reduce any bitterness, peel the cucumber and score down its length with the tines of a fork. (This releases juices that can cause bitterness.)

Serving Suggestions: Cucumbers are excellent in salads, with raw vegetable dips, and as a substitute for lettuce in sandwiches. Seeded cucumber halves can also be filled with chicken, tuna, or seafood salad.

Sliced cucumbers can also be included in Oriental-style stir-fries, light soups, and some other cooked dishes.

ORIENTAL CUCUMBER SOUP

Light and delicately flavored, this soup is particularly nice served as a first course in a Chinese-style meal. The seasonings enhance but don't overpower the subtle taste of the cucumbers.

    1   tablespoon peanut or vegetable oil
    1   small, skinned, boneless chicken breast half, cut into ¼-inch cubes
5 to 6   scallions, white parts cut into ¼-inch lengths and green tops finely
         chopped separately and reserved
    1   small garlic clove, minced
   4½   cups chicken broth
    1   teaspoon soy sauce
    ½   teaspoon red wine vinegar or apple cider vinegar
    ½   teaspoon salt (omit if commercial chicken broth is used)
    ⅛   teaspoon white pepper, preferably freshly ground
    ½   cup cold water
   2½   tablespoons cornstarch
    2   cups peeled, seeded, and diced cucumbers (about 2 medium sized)

Heat the oil in a 3- to 4-quart saucepan or pot over medium-high heat until hot. Add the chicken, white scallion pieces, and garlic and cook, stirring, for 3 to 4 minutes, or until the chicken is opaque and the onion is limp but not browned. Stir in the broth, soy sauce, vinegar, salt (if used), and pepper and bring the mixture to a boil. Lower the heat and simmer, uncovered, for 5 minutes.

In a small bowl or cup, stir the water into the cornstarch until well blended and smooth. Stir the water-cornstarch mixture into the soup and allow it to return to a simmer. Add the cucumber and the reserved chopped scallion tops and continue simmering for 4 to 6 minutes, or until the broth is slightly thickened and clear and the cucumber is just beginning to turn translucent but not soft. Serve the soup immediately; otherwise the cucumber will continue to cook while standing and lose its appealing crispness.

Makes 5 to 6 servings.

CUCUMBER DRESSING

This dressing is good on a salad of greens and crisp raw vegetables.

1   cup peeled, seeded, and coarsely chopped cucumber
¼   cup mayonnaise
¼   cup commercial sour cream
¼   cup plain lowfat yogurt
2   tablespoons chopped fresh chives or scallions (green tops only)
2   tablespoons coarsely chopped fresh parsley leaves
½   teaspoon dried dillweed
½   teaspoon celery salt
½   teaspoon sugar

Combine all the ingredients in a blender or food processor. Blend just until the cucumber is finely chopped. Transfer the dressing to a small bowl or large jar. Cover and refrigerate for at least 1 hour so that flavors can blend. The dressing can be kept for a few days in the refrigerator.

Makes 1¼ cups.

MARINATED CUCUMBER AND CHERRY TOMATO SALAD

Vegetables
2   cups cherry tomatoes
1   medium-sized cucumber

Marinade
⅓   cup vegetable oil
3   tablespoons apple cider vinegar
1½ teaspoons sugar
2   tablespoons chopped fresh parsley leaves
2   finely sliced scallions, green tops only
¼   teaspoon salt
1   teaspoon dried basil leaves (or 2 teaspoons finely chopped fresh basil leaves)
⅛   teaspoon black pepper, preferably freshly ground
⅛   teaspoon dried thyme leaves

To Serve
    Lettuce leaves

Cut each tomato in half. Peel the cucumber and cut it into ½-inch cubes. Set the vegetables aside.

In a medium-sized bowl, combine all the marinade ingredients and mix well. Add the tomatoes and cucumbers to the bowl. Stir to coat them with the marinade. Cover and refrigerate the salad for several hours to allow the flavors to blend. Stir several times to distribute the marinade evenly.

Remove the tomatoes and cucumbers from the marinade with a slotted spoon. Arrange them on a bed of lettuce leaves.

Makes 4 to 6 servings.

GERMAN-STYLE CUCUMBER SALAD

3	tablespoons white wine vinegar or apple cider vinegar
3	tablespoons vegetable oil
1½	teaspoons sugar
2	teaspoons finely chopped scallion, including green top
¼	teaspoon finely chopped fresh dillweed (or ½ teaspoon dried dillweed)
⅛	teaspoon white pepper, preferably freshly ground
⅛	teaspoon powdered mustard
4 to 5	medium-sized cucumbers, peeled and thinly sliced
	About 1 teaspoon salt

For Garnish

Red leaf lettuce or Boston lettuce leaves
2 or 3   sliced radishes (optional)

Combine all the ingredients, *except* the cucumbers and salt, in a cruet or glass jar with a tight-fitting lid. Close the container and shake well. Refrigerate the mixture for several hours to allow the flavors to blend.

Meanwhile, arrange the cucumber slices one layer thick in a flat-bottomed, noncorrosive dish. Sprinkle the first layer of slices thoroughly, but not heavily, with salt. Continue adding layers and salting until all the slices have been used. Weight the cucumbers down with a plate (as if making pickles); cover and refrigerate them for several hours.

Transfer the cucumber slices (which will be almost translucent and limp) to a colander and let them drain. Rinse the cucumbers well under cold water and thoroughly drain them again. Pat the cucumbers dry with paper towels.

Arrange the lettuce leaves on a large salad plate or on individual salad plates. Arrange the cucumbers on the lettuce. Garnish the cucumbers with a few radish slices, if desired. Drizzle the dressing over the cucumbers. Serve immediately, or cover and refrigerate the cucumbers for up to 45 minutes.

Makes 4 to 6 servings.

SAUTÉED TURKEY CUTLETS (OR CHICKEN BREASTS) WITH CUCUMBERS

(main dish)

Many people have never eaten cooked cucumbers, but they are delicious in this elegant yet quick dish.

1	pound thinly sliced turkey breast cutlets (or 4 medium-sized chicken breast halves, skinned and boned)
¼	teaspoon salt
¼	teaspoon black pepper, preferably freshly ground
	About ¼ cup enriched all-purpose or unbleached white flour
1	large or 2 small cucumbers, peeled
⅛	teaspoon dried thyme leaves
⅛	teaspoon dried dillweed
2	tablespoons butter or margarine, divided
1	tablespoon vegetable oil
⅓	cup dry white wine, dry vermouth, or dry sherry
⅓	cup (or to taste) chicken broth or bouillon (reconstituted from cubes or granules)
¼	cup finely chopped fresh parsley leaves

Put the turkey cutlets or chicken breasts between 2 sheets of heavy plastic wrap and pound with a rubber mallet, the flat side of a meat mallet, or rolling pin until the meat is ⅛ inch thick. Be careful not to tear the meat. Lightly sprinkle both sides of each piece with some of the salt and pepper. Then dredge each piece in the flour, shaking off any excess. Set aside.

Cut the cucumber in half lengthwise. Use a melon baller or small spoon to gently scrape all the seeds from each half. Discard the seeds. Cut each half crosswise into ¼-inch-thick "U"-shaped slices. Toss the cucumber slices with the thyme and dill. In a large skillet or sauté pan over medium-high heat, melt 1 tablespoon of the butter. Sauté the cucumber slices until they are cooked through and tender, about 2 minutes. Temporarily transfer the cucumber to a platter.

Add the remaining 1 tablespoon of butter and the oil to the skillet. Lightly brown the poultry pieces on both sides, in batches if necessary. This should take only about 2 minutes for each side; do not overcook the meat or it may toughen. Put the cooked poultry on the platter with the cucumber.

Add the wine to the skillet and heat it to simmering. Then stir in the broth, scraping the bottom of the pan to mix in any browned bits. Bring the sauce to a boil. Return the cooked poultry and cucumbers to the skillet, and continue heating for about 30 to 60 seconds, or until the sauce has thickened slightly. Transfer the poultry and cucumbers to a serving platter. Top with any sauce remaining in the skillet. Sprinkle with the parsley and serve.

Makes 3 to 4 servings.

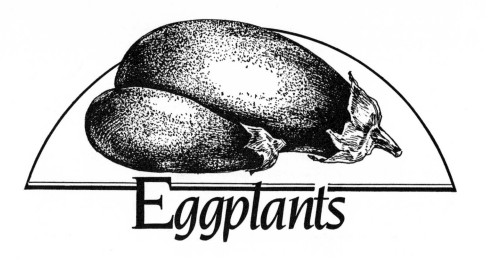

Eggplants

Botanically a large, smooth-skinned berry of the nightshade family, the eggplant has been an important food in the Near and Far East for well over a thousand years. Small varieties were growing in China by the fifth century. Several larger ones were also eaten in ancient India, where wild eggplants can still be found today. The versatile vegetables have likewise been a staple in the Middle Eastern diet for centuries, and one bit of regional folk wisdom states that a woman is not ready for marriage until she knows at least 101 ways to prepare eggplant!

It is thought that this sun-loving plant was introduced into the Mediterranean area by the Arabs during the late Dark Ages. By the sixteenth century it had become known even in northern Europe, although botanists in this region called the species Mala insana, or "mad apple," and believed that eating it caused insanity. English-speaking peoples probably gave it the rather curious name "eggplant" because the varieties they first saw bore egg-shaped fruit.

The Spaniards are credited with bringing the species to the New World. It was already established in Brazil by the mid-1600s and, by the 1900s, numerous varieties were being grown in the U.S., though mostly for ornamental purposes.

Despite its culinary importance in many areas of the world, the eggplant has been slow to catch on in this country. Now, however, per capita consumption is gradually rising as more and more people discover the vegetable's rich, yet subtle, meaty flavor and its usefulness in a wide range of dishes and cuisines. The increased interest has spurred the introduction of several varieties in addition to the large shiny, purplish-black-skinned type normally seen—including miniature "Italian" eggplants about 6 inches long, small, slender delicately flavored "Japanese" eggplants, and even ones with creamy-white and lilac-colored skins.

Availability: The large (7 to 12 inch), pear-shaped or elongated varieties sold in most supermarkets are available in good quantity year round. Although eggplant is a warm-weather crop, a steady supply is furnished from Mexico during the cold months in the U.S. The more exotic types, including the miniature, "Italian" eggplant, the white-skinned, and a small, thin-skinned Japanese variety, are generally only found in produce and ethnic stores and on an irregular basis.

Choosing the Best: Select eggplants that are firm, satiny-smooth, unblemished, and heavy for their size. A wrinkled skin and flabbiness indicate overmaturity (which can cause bitterness), and any pitting or brown spots in the surface signal that deterioration is under way. Unusually lightweight eggplants tend to have spongy or pithy flesh.

When buying standard-sized eggplants, choose those weighing less than a pound unless larger ones are specifically called for in a recipe. The seeds are less prominent and the flesh firmer and slightly sweeter tasting.

Nutritional Value: Eggplant furnishes some potassium and niacin, but is not otherwise a significant source of nutrients. However, it does provide satisfying, rich taste without many calories; 3½ ounces of uncooked eggplant have only about 20.

Storage: Eggplants do not keep well and should be used as soon as possible. They can be held, unwrapped or placed in a paper bag, on a refrigerator shelf for 2 or 3 days. Take care not to bruise the skin surface, as this speeds deterioration and decay.

Preparation and Basic Cooking: Because eggplant is very versatile, basic preparation varies considerably depending on the dish to be made. However, begin by rinsing or wiping off the skin. Unless the eggplant is to be roasted whole, use a sharp knife to cut off the tough stem and leaf cap. In some recipes, the eggplant skin is left intact. However, if instructions call for peeling, do this with a sharp knife or vegetable peeler. The flesh may be cut into strips, chunks, lengthwise or crosswise slices, or hollowed out for stuffing. Although some sources suggest that eggplant should always be "salted" and drained before using, this step is unnecessary in many cases. However, it is a good idea when the flesh is to be fried or sautéed, because it will help prevent the eggplant from absorbing too much fat. To salt eggplant, simply lay the peeled slices or pieces in a colander and sprinkle evenly, but not heavily, with salt. Weight the eggplant down with a plate and let it drain for about 30 minutes so the excess juices can be released; then rinse lightly and pat dry with paper towels.

There are several very basic ways to cook eggplant, including boiling, steaming, broiling, and sautéing. Boil 1-inch cubes in a very small amount of water for 3 to 4 minutes. Steam 1-inch cubes in a steamer basket for 5 to 8 minutes. To broil, brush ½-inch-thick crosswise slices with olive oil and broil under a preheated element for 4 to 6 minutes on one side, and then for 3 to 5 minutes on the other. To sauté eggplant, combine 1-inch (salted and drained) cubes with olive oil or butter and cook, stirring, for 6 to 9 minutes, or until they are just tender.

Simple Serving Suggestions: Plain sautéed, boiled, or steamed eggplant can also be dressed with salt and freshly ground black pepper and a liberal sprinkling of Parmesan cheese. Sautéed eggplant is also delicious cooked along with a little garlic and onion, fresh sliced mushrooms, or sweet green and sweet red pepper and tomatoes. Broiled eggplant is good topped with tomato sauce and mozzarella cheese and baked for a few minutes until the cheese melts.

EGGPLANT AND SWEET PEPPER SAUTÉ

(side dish)

> 2½ tablespoons olive oil
> 1 small garlic clove, minced
> 2 medium-sized scallions, including green tops, chopped
> 1 1-pound eggplant, peeled and cut into 1-inch pieces
> 1 large sweet red pepper, cut into 1-inch chunks
> 4 to 5 medium-sized mushrooms, coarsely sliced
> 2 tablespoons chopped fresh chives (or 1 tablespoon dried chives)
> ⅛ teaspoon salt
> ⅛ teaspoon black pepper, preferably freshly ground
> 1½ tablespoons grated Parmesan cheese

In a large, heavy skillet or sauté pan over high heat, heat the oil until hot but not smoking. Add the garlic, scallions, eggplant, and sweet pepper and cook, stirring constantly, for 4 minutes. Add the mushrooms, chives, salt, and pepper. Lower the heat to medium. Continue to cook, stirring, for about 5 minutes, or until the eggplant is lightly browned and just tender when pierced with a fork. Remove the pan from the heat and stir in the Parmesan cheese.

Makes about 4 servings.

"OVEN-FRIED" BREADED EGGPLANT

(side dish)

Breaded eggplant has long been popular—either as a side dish or incorporated into a main course, such as Eggplant Parmesan. However, eggplant tends to absorb great quantities of oil when it is fried. So we developed this tasty, broiled version of breaded eggplant, which is much lower in fat.

Seasoned Crumb Mixture
2½   cups coarsely cubed fresh or slightly stale bread (may be part or all whole wheat)
2   tablespoons grated Parmesan cheese
1½   tablespoons dried parsley flakes
2   teaspoons instant minced onions
1½   teaspoons dried oregano leaves
1   teaspoon dried basil leaves
½   teaspoon dried marjoram leaves
¼   teaspoon dried thyme leaves
¼   teaspoon garlic powder
¼   teaspoon salt
⅛   teaspoon black pepper, preferably freshly ground

Eggplant
1   medium-sized eggplant (about 1½ pounds)
1   large egg plus 1 large egg white
1   tablespoon vegetable oil

For the seasoned crumb mixture, put all the ingredients in a food processor or blender, and process them until fine crumbs are formed. (This may be done in advance, and the crumbs can be stored for months in the freezer. Thaw them before using.)

Lightly oil or coat a large baking sheet with nonstick vegetable spray.

Cut off and discard the ends of the eggplant. If desired, peel the eggplant (it is not necessary). Cut the eggplant crosswise into ¼-inch-thick slices.

In a small bowl, beat the egg and egg white with the oil. Put the crumbs in another bowl or on a large piece of wax paper or plastic wrap.

Dip each slice of eggplant into the egg mixture; then let the excess egg mixture drip back into the bowl. Coat the eggplant completely with the crumb mixture. Lay out the coated eggplant slices slightly separated on the prepared baking sheet.

Put the filled baking sheet in a preheated broiler about 5 to 6 inches under the heating element and broil the eggplant for about 4 to 7 minutes on each side, or until browned and crunchy. Rotate the pan during broiling, if necessary, for even browning.

Makes about 6 servings.

EGGPLANT ORIENTALE

(side dish)

If you like eggplant and spicy Chinese food, this quick and easy stir-fry is sure to appeal.

Sauce
1	teaspoon cornstarch
¼	cup water
1	tablespoon dry sherry
1	tablespoon soy sauce
1	tablespoon ketchup
1	tablespoon red wine vinegar or apple cider vinegar
2	teaspoons sugar
3 to 4	drops Tabasco sauce

Vegetable Mixture
3	tablespoons peanut or vegetable oil
1	garlic clove, minced
3	scallions, white parts cut crosswise into ½-inch lengths and green tops chopped separately and reserved for garnish
¾	pound unblemished, unpeeled eggplant, cut into 1-inch chunks (about 4 cups)
2	cups small fresh broccoli flowerets
⅓	cup well-drained sliced canned water chestnuts (optional)

Combine the cornstarch and water in a small bowl and stir until well blended and smooth. Stir in all the remaining sauce ingredients and set the mixture aside.

In a large skillet over medium-high heat, heat the oil until hot but not smoking. Add the garlic and white scallion pieces and cook, stirring, for 30 seconds. Add the eggplant and continue to cook, stirring, for 2 minutes longer. Stir in the broccoli and lower the heat to medium. Cook, stirring, for 2 minutes longer. Add the sliced water chestnuts (if used). Briefly stir the sauce mixture and pour it over the vegetables. Cook, stirring constantly, until the sauce is slightly thickened and clear, about 1½ to 2 minutes longer. Transfer the mixture to a serving bowl and sprinkle with the reserved chopped scallion tops.

Makes 4 to 5 servings.

CHEESE-STUFFED EGGPLANT ROLLS

(main dish)

Here's an unusual way to prepare eggplant that is especially appealing and attractive to serve.

1   large eggplant (about 2 pounds)

Filling

1   10-ounce package frozen chopped spinach, thawed and very well drained
1   15-ounce container (about 2 cups) part-skim ricotta cheese (if unavailable, substitute regular ricotta)
4   ounces mozzarella or similar cheese, grated (1 cup packed)
¼   cup grated Parmesan cheese
1   large egg
⅛   teaspoon black pepper, preferably freshly ground
     Pinch of ground nutmeg

Topping

1   8-ounce can tomato sauce
1   teaspoon ground basil leaves
½   teaspoon dried oregano leaves
2   ounces mozzarella or similar cheese, grated (½ cup packed)

To prepare the eggplant, first cut off both of its ends with a sharp knife. Then cut the eggplant lengthwise into very thin (⅛- to ¼-inch-thick) slices, and discard the two outer slices, which will be mostly peel. The remaining slices (about 12) will be almost rectangular in shape, with thin strips of peel along both long sides.

Lightly grease or coat with nonstick spray 1 very large baking sheet or 2 smaller ones. Spread out the eggplant slices in one layer on the baking sheet(s). Bake the slices in a preheated 450-degree oven for about 6 to 8 minutes, or until they are tender and flexible. Remove the eggplant from the oven to cool, and lower the oven temperature to 375 degrees if you will be baking the stuffed eggplant rolls shortly.

While the eggplant is cooling, prepare the filling. Squeeze as much water from the spinach as possible; then put it in a bowl with the remaining filling ingredients. Mix well. Spread some of the filling over each eggplant slice, dividing it evenly. Roll up each slice like a jelly roll, beginning at the narrower of the two unpeeled ends. Place the rolls, seam side down, in a greased or nonstick spray-coated 12- by 7-inch baking dish (or equivalent). (The recipe may be made several hours ahead to this point.)

Just before baking the eggplant rolls, combine the tomato sauce with the basil and oregano, and spoon it over the eggplant rolls; then sprinkle the ½ cup grated cheese on top. Bake the rolls in the preheated 375-degree oven for about 35 to 40 minutes, or until the sauce is hot and bubbly and the eggplant is very tender.

Makes 4 to 6 servings.

Fennel

Also known as Sweet Fennel, Florence Fennel, or *Finocchio* (its Italian name), this vegetable is most popular in the U. S. among those of Italian descent. Recently, it has begun to show up in major supermarkets, where it is often mistakenly labeled "anise," an herb that is a completely different plant. The only thing that fennel and anise have in common is a distinctive, licorice-like flavor. However, fennel has a milder, more subtle taste than anise, particularly when cooked. It is sometimes described as being cool and refreshing when eaten raw.

Unlike anise, which is grown primarily for its seeds and the oil extracted from them, fennel is prized for its swollen, fleshy, enlarged base. This is typically about 3 to 4 inches in diameter. Though the fleshy part of the plant is usually referred to as a fennel "bulb," it is not one in the technical sense, but is actually composed of the overlapping broad bases of several long stalks. The attractive, feathery leaves at the top of the stalks are sometimes used as a garnish.

In medieval times, fennel was credited with many medicinal qualities. It was thought to cure eye diseases and increase a nursing mother's milk supply.

Availability: Formerly available only in Italian or ethnic specialty groceries, fennel has recently begun to appear at many supermarkets. It is in season from October through April.

Choosing the Best: Look for stalks that are firm, crisp, and solid, with a well-formed base or "bulb." The stalks and bulb should be white, cream-colored, or very pale green. If the leaves are still attached, they should be bright green. Avoid fennel that has any discoloration, particularly brown areas, or soft spots. Also, do not choose those bulbs that have cracking or extensive bruising. Yellow foliage is a sign that fennel is past its prime. If the outermost part of the bulb seems to be tough but the fennel otherwise appears to be fresh, it will most likely still be tender inside.

Nutritional Value: Fennel is high in vitamin A and calcium and also contains fair amounts of potassium, phosphorus, and iron.

Storage: Refrigerate fennel in a loosely sealed plastic bag or container or in the refrigerator crisper. It will keep about 3 to 4 days and sometimes up to a week or longer. Do not trim off any leaves or stalks until using the fennel.

Preparation and Basic Cooking: The leaves and slender stalks are edible and are sometimes used in cooking, but the large bulbous base is of primary interest. Cut off the stalks where they meet the bulb. The bottom few inches of each stalk may be used if tender; however, this is sometimes quite tough, as is the upper part of the stalk. Reserve some leaves if desired for a garnish, and discard all tough parts of the stalks.

Slice off the hard bottom of the bulb and discard it. Also peel off and discard any tough outer stalks on the outside of the bulb. Vertically cut the bulb into quarters (or in half, if it is very small), or into slices of any desired thickness. Alternatively, the base may be sliced horizontally; however, the pieces will separate like the rings of an onion when cooked.

Fennel may be eaten raw or cooked in any of a number of ways, such as simmering, steaming, braising, sautéing, or frying. It should be cooked just until tender. Simmered fennel takes about 8 to 20 minutes to cook, depending on the size of the pieces and the toughness of the fennel. It should be tender when pierced in the center with a fork.

Simple Serving Suggestions: Cut raw fennel into julienne strips, crosswise into rings, or dice, and use in a tossed salad. Serve cooked fennel with butter, garnished with some of the fresh leaves from the plant (if available), and possibly a sprinkle of grated Parmesan cheese.

BRAISED FENNEL WITH PARMESAN CHEESE

(side dish)

When fennel is in season, this makes an unusual yet easy dish to complement fish, roasted fowl, or other meat. It is a nice way to get to know the delicious taste of fennel.

2	medium-sized fennel bulbs
1	tablespoon olive oil
1	small garlic clove, finely minced
¼	cup chicken broth or bouillon (reconstituted from cubes or granules) or water
1 to 1½	tablespoons grated Parmesan cheese (to taste)
	About 1 tablespoon chopped fennel leaves or fresh parsley leaves

Slice off and discard the hard base of each fennel bulb. Cut off the fennel stalks about ½ inch above the point where they meet the bulbs. (Reserve some of the leaves, if desired, for a garnish; discard the remainder of the trimmings.) Vertically cut each fennel bulb crosswise through the stalks into ½-inch-thick slices (or cut each bulb vertically into quarters, to produce wedge-shaped pieces.)

In a large skillet over medium-high heat, heat the oil; then add the fennel and sauté until it is lightly browned on all sides, about 3 to 5 minutes. Add the garlic and broth and immediately cover the skillet. Lower the heat and simmer the fennel pieces, turning them often, for about 7 to 15 minutes, or until they are tender in the center when pierced with a fork. (The exact time will depend on the size of the pieces and the toughness of the fennel.)

Remove the cover, raise the heat slightly, and boil the liquid down, stirring to scrape up any browned bits in the bottom of the skillet, until most of the liquid has evaporated, and only a light sauce remains. Sprinkle the Parmesan cheese on top of the fennel, cover the skillet, and heat the fennel briefly just until the cheese begins to melt. Sprinkle the fennel leaves on top as a garnish.

Makes about 4 servings.

FENNEL WITH FRESH TOMATO AND GARLIC

(side dish)

3   medium-sized fennel bulbs
1½   tablespoons butter or margarine
1   tablespoon olive oil
1   small garlic clove, minced
⅓   cup water
¼   teaspoon salt
⅛   teaspoon black pepper, preferably freshly ground
1   large tomato, peeled, seeded, and chopped

If the fennel bulbs are purchased untrimmed, trim off all but the bottom 5 to 6 inches of the bulbs. Also trim a thin slice from the bottom of each bulb. Pull or cut off any dry or tough outer stalks and discard them. With a sharp knife, cut each bulb vertically into 3 or 4 ¼-inch-thick slices. Rinse the slices and drain them well in a colander.

Heat the butter and olive oil in a large skillet over medium-high heat until the butter melts. Add the garlic and cook, stirring, for 1 minute. Stir in the water, salt, and pepper. Lay the fennel slices in the skillet. Top them with the chopped tomato. Bring the mixture to a boil; then lower the heat. Gently simmer the fennel, uncovered, for 10 minutes. Turn the slices over and continue cooking for 10 to 12 minutes longer, or until the slices are tender when pierced in the center with a fork (the outer parts of the slices will still be slightly firm). If necessary, add a bit more water to the skillet to prevent it from boiling dry. Spoon the remaining pan liquid and tomato over the fennel slices and serve.

Makes about 4 servings.

Kale

Kale is a botanically primitive, "headless" member of the cabbage family. It appears in today's gardens in much the same form as it did several thousand years ago. The ancient Romans enjoyed several different kinds of kale, and this crinkly leafed green was a part of the Anglo-Saxon diet as well.

Perhaps kale has changed so little over time simply because horticultural fiddling seemed unnecessary. In addition to being among the most vigorous, prolific, and easy-to-grow vegetables, this uncomplicated plant is resistant to cold; simple to harvest, store, and prepare; and rich in vitamins and minerals.

Although kale tends to be overlooked and underappreciated in the U.S., the pleasantly pungent green vegetable does have its enthusiasts elsewhere. The Scottish and Irish are the world's biggest kale eaters, and it is popular in northern Germany, as well. In fact, the regional dish in the city of Bremen is kale with little sausages, and the residents refer to the first frosty days of autumn as "kale time."

Availability: Kale is a cool-weather vegetable, and many gardeners seem to agree with the north Germans that light frost brings the leaves to the peak of flavor. Perhaps as a result, it is most abundant and inexpensive from November through March. However, it can be found in many markets year round.

Choosing the Best: Fresh, top-quality kale leaves are always springy to the touch, and, depending on the variety, moderately to very curly. One commercially sold variety, Blue Kale, has fairly crinkly leaves and a rich, bluish-green color. A second type, Scotch Kale, is curly to the point of frizziness and deep gray-green. Both varieties should look crisp and have good color. Avoid greens that are limp, shriveled, yellowed, or brown. If the leaves are packed in plastic bags, also be wary of excessive condensation inside. This can signal deterioration.

As a rule, the smaller the leaves, the more tender and mild the greens will be.

Nutritional Value: Kale is very rich in vitamins A and C. It is also a good source of B vitamins, calcium, potassium, and other minerals.

Storage: Store kale loosely packed in a plastic bag in the coolest part of the refrigerator. It should be used within 3 or 4 days, and preferably sooner. Do not wash before storing.

Preparation and Basic Cooking: Wash and drain kale in a colander. Strip or cut the leafy portions of the greens from the coarse ribs and stems (discard the ribs and stems); tear the leaves into bite-sized pieces. Boil kale in a small amount of water for 15 to 18 minutes, or steam in a steamer basket for 20 to 25 minutes. Be careful to cook the leaves only until they are tender, but not mushy. Also serve immediately; the greens are delicious when just cooked but sometimes develop a strong taste upon standing.

Simple Serving Suggestions: Enliven kale with a bit of salt, pepper, and butter. To tame the slightly sharp taste of kale, simmer in a bit of beef bouillon instead of water. To accent its pungent quality, add a dash of lemon juice just before serving, or pass a cruet of apple cider vinegar at the table. As with other members of the cabbage family, kale goes well with pork.

VELVETY KALE SOUP

The list of ingredients doesn't begin to suggest just how tasty this easy soup is. It's very healthful, too.

1	pound fresh kale
2	tablespoons butter or margarine
5 to 6	scallions, including green tops, coarsely chopped
1	large garlic clove, minced
3	cups coarsely cubed, peeled potatoes
1	quart chicken stock or broth
2	tablespoons dried chopped chives
1	teaspoon lemon juice, preferably fresh
½	teaspoon salt (omit if canned chicken broth is used)
⅛	teaspoon black pepper, preferably freshly ground
	Pinch of ground mace
1 to 1½	cups lowfat or whole milk (approximately)

Wash the kale and trim off the stems and any coarse ribs. Tear the leaves into bite-sized pieces. Rinse again and set aside in a colander to drain thoroughly.

Melt the butter in a large pot over medium-high heat. Add the scallions and garlic and cook, stirring, for 4 to 5 minutes, or until they are limp. Add the kale and cook, stirring, until the leaves begin to wilt and are coated with butter. Stir in all the remaining ingredients, *except* the milk, and bring the mixture to a boil. Lower the heat and simmer, uncovered, for 13 to 15 minutes, or until the potatoes and kale are tender; stir occasionally to prevent the potatoes from sticking to the bottom of the pot.

Stir 1 cup of milk into the mixture. In batches, transfer the mixture to a blender and purée until it is completely smooth. Return the puréed mixture to the pot and reheat until piping hot. If the soup is very thick, gradually add more milk until the desired consistency is obtained.

Makes 5 to 7 servings.

KALE AND CABBAGE SALAD

The pungent kale leaves and tangy dressing give this wonderful salad its unusual flavor. It should be refrigerated for at least an hour before serving. But it tastes even better if made the day before it's needed.

Vegetables
3   cups (lightly packed) kale leaves, stems and coarse midribs removed
3   cups coarsely shredded green cabbage
¼   cup finely chopped red onion (optional)
1   large tomato, cut into ½-inch cubes

Dressing
1   tablespoon dry white wine or dry sherry
1   teaspoon Dijon-style mustard
¼   cup mayonnaise
¼   cup commercial sour cream
¼   teaspoon curry powder
¼   teaspoon salt
⅛   teaspoon black pepper, preferably freshly ground

Tear the kale leaves into small pieces. Combine them with the other vegetables in a medium-sized bowl.

In a small bowl, mix together all the dressing ingredients. Add the dressing to the vegetables. Toss to coat the vegetables evenly. Cover the salad and refrigerate for at least 1 hour before serving.

Makes 6 to 7 servings.

KALE WITH RED ONION

(side dish)

1   pound fresh kale, thoroughly rinsed and drained
1   small red onion, cut crosswise into ⅛-inch-thick slices
1   tablespoon butter or margarine
⅓   cup beef broth or bouillon (reconstituted from cubes or granules)
½   teaspoon lemon juice, preferably fresh
⅛   teaspoon black pepper, preferably freshly ground
    Pinch of ground nutmeg

Trim off and discard the stems and coarse ribs from the kale leaves, using a sharp knife. Tear any large leaves in half. Separate the onion slices into rings. Combine the onion rings and butter in a 6-quart saucepan or pot over medium-high heat.

Cook, stirring occasionally, for 4 to 5 minutes, or until the onion is limp but not browned. Remove the onion from the pan with a slotted spoon and reserve it in a small bowl.

Add the broth, lemon juice, pepper, and nutmeg to the pan. Bring the mixture to a simmer. Add the kale, stirring for several minutes, until the leaves begin to wilt and are all moistened with the pan liquid. Cover the pan with a tight-fitting lid. Cook the kale for 12 to 16 minutes, or until the leaves are almost tender; stir and check once or twice to make sure the pan isn't boiling dry, and add a bit of water if necessary. When the kale is almost tender, return the red onion to the pan and stir until incorporated. Cook for 1 to 2 minutes longer, or until the onion is reheated and the kale is just tender. Serve immediately; kale sometimes develops a strong taste if allowed to stand.

Makes 4 to 5 servings.

KALE AND POTATO CASSEROLE

(side dish)

2   pounds boiling potatoes (about 7 medium-sized), peeled and cut into
     ¼-inch-thick slices
1   pound fresh kale, stems and midribs removed
1   tablespoon instant minced onions
     Generous ½ teaspoon salt
     Black pepper to taste, preferably freshly ground
3   tablespoons cold butter or margarine
¾   cup whole milk

In a medium-sized saucepan, combine the potatoes with enough water to cover them. Bring to a boil over medium-high heat and cook, uncovered, for 5 minutes. Turn out the potatoes into a colander and drain well.

Blanch the kale by putting it in a large colander and pouring several quarts of boiling water over it; let it drain well.

Lay half the potato slices in a 4- to 5-quart Dutch oven. Sprinkle a third of the onion over the potatoes. Sprinkle the potatoes with about a third of the salt and black pepper to taste. Dot the potatoes with 1 tablespoon of the butter. Lay the drained kale over the potatoes; press it down with your hands to form a compact layer. Sprinkle the kale with another third of the onion, salt, and black pepper to taste. Dot the kale with another tablespoon of the butter. Top the kale with the remaining half of the potatoes. Sprinkle the potatoes with the remaining onion, salt, and pepper to taste. Dot with the remaining tablespoon of butter. Pour the milk over the casserole.

Cover the casserole and bake in a preheated 375-degree oven for 45 to 50 minutes, or until the potatoes are just tender and most of the milk has been absorbed.

Makes 5 to 6 servings.

MIXED VEGETABLE STUFFING OR CASSEROLE

(side dish)

Here's an unusual yet tasty stuffing for Thanksgiving turkey. Unlike bread-based stuffings, which tend to be very heavy, this one is light and even colorful. The combo is equally good baked in a casserole. (Note: A food processor can save time in shredding and chopping the vegetables.)

3	tablespoons butter or margarine
1	large onion, finely chopped
2	medium-sized, thin-skinned potatoes, scrubbed and shredded (or 1 large baking potato, peeled and shredded)
2	medium-sized turnips, peeled and shredded
4	medium-sized carrots, shredded
½	small green or Savoy cabbage, shredded
1½	cups loose-pack frozen corn kernels
1	10-ounce package frozen chopped kale, thawed and drained
¼	cup finely chopped fresh parsley leaves
1	teaspoon dried basil leaves
¼	teaspoon dried thyme leaves
½	teaspoon salt
⅛ to ¼	teaspoon black pepper, preferably freshly ground
½	cup chicken broth or bouillon (reconstituted from cubes or granules) or vegetable stock (page 86)
2	tablespoons imitation bacon bits (optional)

In a large pot or Dutch oven over medium-high heat, melt the butter; then cook the onion until it is tender but not browned. Add the potatoes, turnips, carrots, and cabbage and stir constantly for about 1 minute. Then add the remaining ingredients, cover the pot tightly, and turn the heat to low. Steam the vegetable mixture for about 5 minutes, stirring occasionally, or until the vegetables are well mixed and heated through. (If they seem to be very dry and are sticking to the bottom of the pot, add a few tablespoons of water.)

Remove the vegetable mixture from the heat and cool it slightly. Use a slotted spoon to transfer the vegetable stuffing into the body and neck cavities of a turkey or other fowl; then roast the poultry as desired.

Or transfer the mixture to a greased or nonstick spray-coated 2-quart casserole dish, and bake, covered, in a 325- to 350-degree oven for about 30 minutes. Then remove the cover and bake for about 20 to 30 minutes longer, or until all the vegetables are tender.

Makes enough vegetable stuffing for an approximately 14-pound turkey or about 8 side-dish servings.

SAVORY PORK AND VEGETABLES ONE-POT DINNER

(main dish)

2	tablespoons vegetable oil
1	3¼- to 3½-pound boneless fresh pork shoulder roast, trimmed of excess fat
2	onions, coarsely chopped
1	large celery stalk, chopped
1	cup coarsely chopped green cabbage
½	cup peeled and coarsely chopped rutabaga
2½	cups water
1	small bay leaf
⅛	teaspoon dried thyme leaves
1	teaspoon salt
⅛	teaspoon black pepper, preferably freshly ground
⅓	cup pearl barley
4 to 5	medium-sized carrots, coarsely sliced
3 to 4	medium-sized potatoes, peeled and cut into eighths
¾	pound fresh kale, stems and midribs removed, torn into bite-sized pieces

Heat the oil in a 5- to 6-quart pot or Dutch oven over medium-high heat until moderately hot. Add the pork roast and brown on all sides, turning frequently. Remove the roast from the pot and set it aside. Add the onion, celery, cabbage, and rutabaga and cook, stirring, for 4 to 5 minutes, or until the vegetables are limp; scrape up any browned bits on the pot bottom during cooking. Stir in the water, bay leaf, thyme, salt, and pepper. Return the pork roast to the pot and bring the mixture to a boil. Lower the heat, cover the pot tightly, and simmer gently for 30 minutes.

Stir the barley into the pot. Cover and simmer gently for 30 minutes longer. Stir all remaining ingredients into the pot and continue cooking, covered, for 30 to 40 minutes, or until the barley, carrots, and roast are just tender. Cut the roast into thick slices and serve in individual soup plates along with the vegetables. The recipe may also be made ahead and then refrigerated and reheated at serving time. In this case, thin the vegetable mixture by stirring in ½ to ⅔ cup water before reheating.

Makes about 6 servings.

Kohlrabi

Hardy and cool-loving, kohlrabi is one of the few vegetables to originate in northern Europe. Its name is a compounding of two German words—*kohl*, which means cabbage, and *rabi*, which means turnip. This may reflect the fact that the trimmed bulbous portion of the plant looks rather like a cabbage-green or greenish-violet turnip.

Although a relative of the turnip, the kohlrabi is not one. In fact, it isn't even a root vegetable. The root-shaped globe is actually an enlarged, succulent part of the plant's main stem! This is topped with slightly ruffled leaves, which protrude upward on long, thin stalks.

While its exact origins are uncertain, experts speculate that kohlrabi may be a fairly recent descendant of a nonheading, thick-stemmed wild cabbage that grows along the European and English Channel coasts. The first recorded mention of kohlrabi, or "stem-turnip" as it is sometimes called, was made in the mid-1500s by a botanist who noted that it had just been introduced into Italy.

One of the many vegetables in the large cruciferous, or cabbage, family, kohlrabi has firm, crisp flesh and a very delicate, slightly "turnipy" flavor. When sliced or cubed and cooked, it seems more elegant than turnip, however, since the pale greenish white flesh becomes appealingly translucent and glossy looking. Like turnip, kohlrabi may also be eaten raw.

Two varieties of kohlrabi are grown commercially in the United States. The pale green "White Vienna" type accounts for about two thirds of the supply, and the violet-tinged "Purple Vienna" accounts for the rest. The two kinds taste nearly identical.

Availability: Kohlrabi is not yet widely known in this country and is difficult to find in many supermarkets. Produce stores and greengrocers catering to families of German and other Central European extraction are most likely to stock the vegetable. Quantities are largest in June and July and September and October.

Choosing the Best: Always choose the smallest kohlrabi available (preferably 2 to 3 inches in diameter or less), as these will be the most tender and least likely to be woody. Also look for ones with smooth, unblemished skins. If the vegetables are being sold untrimmed, choose those with crisp stems and leaves. When in good condition, these can be used as greens in soups.

Nutritional Value: Kohlrabi is a good source of potassium and vitamin C and also contains some magnesium. It is also low in calories; a ¾-cup serving of boiled and drained kohlrabi has only about 25.

Storage: Placed in a loosely closed plastic bag and refrigerated, kohlrabi will keep well for up to a week. When the vegetables are purchased untrimmed, do not wash or trim until just before using.

Preparation and Basic Cooking: To prepare untrimmed kohlrabi, rinse well and use a sharp knife to cut away the root and all leaves protruding from the bulb-like part. (The leaves and any very tender leaf stems can be reserved and used in soups and stews if desired.) Using a vegetable peeler or knife, remove every bit of tough, fibrous skin from the trimmed kohlrabi. It can then be sliced, diced, or shredded and used raw. To cook kohlrabi, cut the flesh into ¼-inch-thick slices and boil in a small amount of water or chicken broth for about 10 to 15 minutes. Or steam kohlrabi slices for about 15 to 20 minutes, or until they are tender.

Simple Serving Suggestions: Use uncooked, diced, or grated kohlrabi to add a pleasant crunch and radish-like zest to salads. Season plain-cooked kohlrabi with butter, salt, and pepper, and perhaps a dash of lemon juice. Dress up cooked, drained kohlrabi by adding a tablespoon or two of cream to the saucepan shortly before serving, or a bit of sour cream combined with a sprinkling of dillweed or pinch of tarragon; then reheat to hot but not boiling. The vegetable is also good napped in a Mornay or light mustard sauce.

KOHLRABI-DILL SOUP

 2 tablespoons butter or margarine
 1 large onion, diced
 3 small kohlrabi, peeled and diced
 4 cups chicken broth or bouillon (reconstituted from cubes or granules)
 1½ tablespoons chopped fresh dillweed (or 2 teaspoons dried dillweed)
 ⅛ teaspoon white pepper, preferably freshly ground
 ¼ cup cold water
 1 tablespoon cornstarch
 ⅛ teaspoon salt (or to taste)
    Commercial sour cream for garnish (optional)

Melt the butter in a large saucepan over medium-high heat. Add the onion and cook, stirring, for 3 to 4 minutes, or until it is limp. Add the kohlrabi and cook, stirring, for 3 to 4 minutes longer. Stir in the broth, dillweed, and pepper. Bring the mixture to a simmer. Lower the heat and simmer, uncovered, for 9 to 12 minutes, or until the kohlrabi is just tender.

Stir together the water and cornstarch until well blended. Add the mixture to the saucepan and cook, stirring, about 1 minute longer, or until the soup is slightly thickened and clear. Add the salt. Garnish each individual serving with a small dollop of sour cream, if desired.

Makes 4 to 5 servings.

KOHLRABI GRATIN

(side dish)

The perfect choice when you're looking for a dish that's not only delicious and elegant, but a bit unusual.

5	medium-sized kohlrabi (about 1¼ pounds), trimmed, peeled, and cut into ¼-inch-thick slices
1¼	cups chicken broth or bouillon (reconstituted from cubes or granules)

Sauce and Cheese

	Reserved kohlrabi cooking liquid
1	tablespoon butter or margarine
2	tablespoons enriched all-purpose or unbleached white flour
2	teaspoons Dijon or Dijon-style mustard
¼	cup light cream or half-and-half
⅛	teaspoon white pepper, preferably freshly ground (optional)
1	ounce mild Cheddar cheese, grated or shredded (¼ cup packed)

Combine the kohlrabi slices and broth in a medium-sized saucepan and bring to a boil over medium-high heat. Lower the heat and simmer, uncovered, for 10 minutes. Remove the pan from the heat. With a slotted spoon, remove the kohlrabi slices from the pan; reserve the cooking liquid in a measuring cup. Arrange the kohlrabi slices attractively in a lightly greased 10- or 11-inch-oval and 2-inch-deep gratin pan or similar shallow casserole.

Add enough water to the reserved kohlrabi cooking liquid to yield ¾ cup; set aside.

Melt the butter in a small heavy saucepan over medium-high heat. Stir in the flour until well blended. Cook the mixture, stirring, for 2 minutes. Stir in the mustard until well mixed. Gradually stir in the cooking liquid-water mixture until the ingredients are well blended and smooth. Bring the mixture to a simmer and cook, stirring, for 1 minute longer. Stir in the cream and pepper (if used) until the sauce is well blended. Remove the sauce from the heat. Pour the sauce evenly over the kohlrabi slices. Sprinkle the grated cheese evenly over the sauce. Bake the gratin in a preheated 375-degree oven for 20 to 25 minutes, or until the sauce is bubbly and the kohlrabi is tender.

Makes 4 to 5 servings.

Lettuce

Lettuce is thought to have originated from a weed that is still found in its wild state in America, Europe, and Asia. Today there are over 200 cultivated varieties around the world.

The ancient Persian kings enjoyed lettuce, and it was a popular vegetable in Ancient Rome. Hippocrates, the father of modern medicine, thought it had health-promoting properties.

The first lettuce was of the loose-leaf type. Firm-headed varieties were developed later, but were well known in Europe by the sixteenth century.

Lettuce was cultivated in the New World soon after Columbus' arrival. Until it became commercially important in the twentieth century, it was grown extensively in home gardens.

There are five main types of lettuce:

Butterhead is a small, loose-leafed head lettuce, of which *Boston* and *Bibb* are the most important commercial varieties. The leaves are pliable, delicate in flavor and bruise easily. The outer leaves are light green, the inner ones light yellow.

Crisphead is commonly called "iceberg" lettuce. Because it is disease-resistant and ships well, it is more widely available than other types. Although rather bland in flavor, it does have a pleasingly crisp texture. The leaves form compact, solid heads by overlapping one another in a smooth regular pattern. The heads are usually at least six inches in diameter.

Leaf or bunching lettuce does not form heads. Instead the leaves are attached to a short central stem, and only those at the center of the plant overlap to any extent. The leaves are tender and may be curled or somewhat smooth. There is considerable color variety, ranging from light green to reddish. Leaf lettuce has a fresh and delicate flavor and is popular with home gardeners. It is also the chief variety produced in greenhouses.

Romaine or cos has long, stiff, upright, crisp leaves. Those on the outside are dark green. Those near the center are greenish-white. The leaves of some varieties curve inward at the tips. The head is characteristically upright and loaf-shaped.

Stem lettuce is different from other varieties because the enlarged stem, and not the leaf, is the edible portion. "Celtuce" is the only type widely available in the U. S. It is used in Chinese cooking and can also be peeled and eaten raw.

Availability: Iceberg lettuce, grown chiefly in California, is available all year round, as is romaine. More tender and perishable types, such as Bibb and Boston, are less readily available but are sold year round in some areas.

Choosing the Best: With all varieties, select fresh looking leaves with no brown discolorations. Color should be bright. Avoid faded, wilted, or soft heads.
   Iceberg should be firm and heavy for its size. Extreme large heads may be overmature and bitter.

Nutritional Value: Lettuce has some vitamin C and A but cannot be considered a major source of either. It is high in fiber and low in calories. Three and a half ounces have 13 to 18 calories.

Storage: Lettuce is highly perishable and should be refrigerated as soon as possible after purchase. Some authorities recommend washing it before storage. Others suggest waiting until just before it is used. In any case, rinse under running water and dry well. For romaine, Boston, Bibb, or leaf lettuce, separate individual leaves during rinsing and dry well. Store all varieties well wrapped in a plastic bag in the vegetable crisper. Iceberg can be held for up to 5 days. The more tender types will keep well for only 2 or 3 days.

Preparation: Although lettuce is the mainstay of many salads, there is considerable controversy about whether the leaves should be cut or torn. Cutting, however, does have a tendency to brown the leaves at the cut edge. With romaine, remove coarse ribs and discard.
   Since lettuce wilts easily, it is best to dress salads just before serving.

Simple Serving Suggestions: Toss lettuce and other salad fixings with a homemade dressing. For instance, combine ¼ cup vegetable oil, 2 tablespoons herb vinegar (such as basil, tarragon, or mixed herb), ½ teaspoon sugar, ¼ teaspoon celery salt, and ⅛ teaspoon freshly ground black pepper in a jar with a tight-fitting lid. Shake to mix well.

Lettuce leaves are also used as a garnish for chicken, egg, vegetable, seafood, and gelatin salads.

LETTUCE AND GREEN PEA SOUP

This makes a nice beginning to a fancy meal. The soup has an attractive pale green color and a rich, delicate flavor.

1½ tablespoons butter or margarine
2 tablespoons coarsely chopped onion
3 tablespoons enriched all-purpose or unbleached white flour
3 cups chicken broth or bouillon (reconstituted from cubes or granules)
⅛ teaspoon white pepper, preferably freshly ground
   Pinch of dried tarragon leaves
3 cups coarsely shredded green leaf lettuce or Boston lettuce leaves
1½ cups loose-pack frozen green peas (preferably baby peas)
1 cup whole milk or half-and-half

Melt the butter in a large saucepan over medium-high heat. Add the onion and cook, stirring occasionally, for 3 to 4 minutes, or until the onion is limp. Stir in the flour until well blended and smooth. Cook, stirring, for 2 minutes longer. Gradually add the broth; then add the pepper and tarragon, stirring vigorously until the mixture is smooth and well blended. Stir in the lettuce and lower the heat. Simmer the mixture, uncovered, for 10 minutes.

In batches, if necessary, transfer the mixture to a blender and blend on medium speed until it is completely puréed and smooth. Return the puréed mixture to the saucepan and add the peas. Bring the mixture to a simmer and cook for 2 to 3 minutes, or until the peas are tender. Stir in the milk and heat the mixture to piping hot.

Makes 4 to 6 servings.

MIXED GREEN SALAD

Dressing
- ½ cup commercial buttermilk
- ⅓ cup mayonnaise
- Scant ¼ teaspoon powdered mustard
- ⅛ teaspoon black pepper, preferably freshly ground
- ⅛ teaspoon onion powder
- Pinch of ground thyme
- Pinch of ground marjoram
- Scant ¼ teaspoon salt

Salad
- 7 to 8 cups lettuce leaves, washed and torn into small pieces (Use a combination of at least two of the following: romaine, red, Boston, and iceberg lettuce.)
- 1 medium-sized cucumber, peeled and sliced
- 2 celery stalks, thinly sliced
- 1 sweet red or green pepper, cut into ½- by 1-inch slices

In a small bowl, combine all the dressing ingredients. Stir with a wire whisk until the dressing is thoroughly combined. Cover and refrigerate for at least 30 minutes.

Just before serving, combine all the salad ingredients in a medium-sized salad bowl. Add the dressing and toss to coat the vegetables well.

Makes 5 to 6 servings.

TOSSED SALAD WITH ZESTY GARLIC DRESSING

Salad

3   cups torn red leaf lettuce or green leaf lettuce leaves
2   cups torn romaine lettuce leaves
1   cup torn escarole leaves
1   cup coarsely shredded red cabbage
1   cup halved cherry tomatoes
½   cup thinly sliced red radishes
⅓   cup thinly sliced fresh mushrooms

Dressing

¼   cup vegetable oil
3   tablespoons red wine vinegar
1   small garlic clove, minced
1   tablespoon finely chopped scallion, including green top
1   tablespoon grated Parmesan cheese
1   teaspoon Dijon or Dijon-style mustard
1   teaspoon sugar
½   teaspoon celery salt
⅛   teaspoon black pepper, preferably freshly ground

Combine all the salad ingredients in a large salad bowl and toss until well mixed.

Combine all the dressing ingredients in a glass jar with a tight-fitting lid **or a** cruet. Shake until the ingredients are thoroughly blended.

Pour the dressing over the salad. Toss until the vegetables are lightly coated with the dressing.

Makes 5 to 7 servings.

GREEK SALAD IN PITA POCKETS

Although this salad is quite tasty on its own, it becomes a real treat when served inside loaves of flat, round pita bread. In fact, this dish would be perfect for a picnic or brown bag lunch if you pack the salad, dressing, and pita bread separately, and just put it all together at mealtime.

The salad is particularly quick to prepare if you use a food processor or rotary food slicer fitted with a thin slicing blade to slice the vegetables.

Salad
1   medium-sized head romaine or iceberg lettuce
    About 12 medium-sized red radishes
1   medium-sized sweet green pepper
1   medium-sized red onion
1   medium-sized cucumber, peeled if desired
    About 12 cherry tomatoes, halved, or 4 medium-sized ripe tomatoes, each cut into 6 wedges
    About 24 Greek-style ripe olives (or plain ripe [black] olives)
    About 1 pound feta cheese

Dressing
½   cup good-quality olive oil
¼   cup red wine vinegar
½   teaspoon dried dillweed
½   teaspoon dried oregano leaves
⅛   teaspoon black pepper, preferably freshly ground
    Pinch of salt

To Serve
About 8 large loaves whole wheat or white pita bread

Very thinly slice (do not chop) the lettuce, radishes, green pepper, red onion, and cucumber. Put all the vegetables in a large bowl. Add the tomatoes, olives, and feta cheese; then toss the salad gently. Cover and refrigerate the salad until serving time.

For the dressing, put all the ingredients in a small jar, cover tightly, and shake well. Just before serving, shake the dressing again, and pour it over the salad. Toss gently.

Cut a small piece from the top of each pita loaf, so that it forms a large pocket. (Or cut each loaf in half to form two, smaller, semicircular pockets.) Use tongs or a spoon to stuff the salad into the pita pockets.

Makes about 8 servings.

Mushrooms

Gastronomically speaking, the mushroom may be considered aristocratic, but botanically, it is our lowliest vegetable of all. It is a fungus of the Thallophyta family, a primitive plant group that also includes algae and lichens. Like other fungi, the mushroom is structurally simple, lacking leaves, a true stem, and chlorophyll. The umbrella-shaped edible portion, or stemmed cap, is actually the plant's fruit; the rest of the mushroom is root-like in appearance and grows underground.

There are many different species and varieties of mushrooms (both edible and inedible), but only one kind, *Agaricus bisporus,* is cultivated on a large scale in the United States. This type is generally plump and dome-like, and may be white, cream, or tan, depending on the particular strain. All three strains have the same subtle, nutty-sweet taste. Sometimes, specialty shops also sell other, more exotic cultivated mushrooms and wild varieties, including morels, shiitake, chanterelles, cepes, and enokidake. Each of these has its own distinctive appearance and flavor, and all are sought after by gourmet cooks.

Mushrooms have been eaten and considered a delicacy for at least several thousand years. They were first described as the "mysterious night-growing" vegetables in Egyptian hieroglyphics, and the pharaohs pronounced them food "fit only for kings." The Greeks and Romans likewise reserved mushrooms for consumption by the upper classes. Julius Caesar even passed stringent laws specifying who could enjoy them and who could not.

Of course, there are also references throughout history to the many poisonous mushroom varieties. As the Great Herball book of 1526 noted, some types were "deadly and sleeth them that eateth of them, and they be called tode stoles."

Nevertheless, edible mushrooms continued to be prized, and by the 1600s, the French had begun cultivating them to keep up with the local demand. Most of the farming took place in caves that had once been mined for building stone. In 1867, one cave in Mery was said to contain 21 miles of beds producing 3,000 pounds of mushrooms a day.

Today in the U. S., mushroom farming is an important and still-growing industry. (Pennsylvania's Kennet Square is the nation's "mushroom capital," and the state produces nearly half of the American total.) Most of these fast-growing vegetables are raised in tiers of beds in special windowless, climate-controlled cinderblock or cement houses.

Once a bed is planted, it takes about six weeks for the first growth of tiny mushrooms, or "pins," as commercial growers call them, to appear. Unless the temperature in the house is kept quite low, the pins will be ready for harvesting in a matter of days (or even hours)! The mature mushrooms are plucked by hand from their base with a twisting motion, then trimmed and packed for shipping and sale.

Availability: Since mushrooms are grown indoors in a controlled environment, a steady supply is available year round. However, the quantity does tend to be lowest in August and rises somewhat during the winter months. In addition, canned mushrooms are readily available in most American supermarkets, although they taste quite different from fresh. Some exotic, imported mushroom varieties are also sold in this country dried, and these need to be soaked before use.

Choosing the Best: When selecting fresh mushrooms, look for ones with smooth, firm, unblemished skins. Also choose mushrooms with fairly light-colored, closed or nearly closed caps; darkening and spreading of the gills, or folds, underneath the caps suggests overmaturity. Wrinkles and softness almost always indicate deterioration. In addition, avoid mushrooms that are caked or flecked with soil, as these will be rather difficult to clean.

Nutritional Value: Mushrooms are a very good source of potassium and also contain some phosphorus, niacin, and folic acid. Additionally, they are also low in calories; a 3½-ounce serving of raw mushrooms has only 28. Although mushrooms are sometimes recommended to dieters, it is best not to eat large quantities of them because they contain chemicals called hydrazines, which have proved to be carcinogenic in some laboratory animals.

Storage: Fresh mushrooms are quite perishable and should be stored in the refrigerator. They need to be packed in a moisture- and air-permeable container or open plastic bag that permits good ventilation, but at the same time prevents them from drying out. Do not clean mushrooms until just before using. They will keep for about 3 or 4 days.

Basic Preparation and Cooking: Mushrooms can become waterlogged if soaked, but in most cases, they do need a thorough rinsing. (If they look completely clean, simply wipe them off with damp paper towels.) Slice off the ends of the stems if they seem dry, tough, or discolored. Mushrooms are now ready to be used raw or cooked.

To sauté mushrooms, coarsely slice and combine them with a tablespoon or two of butter in a large skillet or sauté pan over medium-high heat. Sauté, stirring, for 3 to 4 minutes, or until they are just tender; do not overcook or the mushrooms will exude their juices.

Mushroom caps can also be broiled. Toss them with a little oil or brush with butter and place them, undersides facing down, under a preheated broiler element for 2 to 3 minutes. Turn the caps over and broil for about 2 minutes longer, or until they are cooked through.

Large whole mushroom caps (2 to 3 inches in diameter) are likewise good baked. Arrange the caps, undersides facing up, in a flat baking dish. Brush or drizzle the caps lightly with butter and perhaps a tablespoon or two of cream. Cover the dish and bake in a preheated 375-degree oven for 12 to 15 minutes, or until the caps are tender but not soft when pierced with a fork.

Simple Serving Suggestions: Sautéed, broiled, or baked mushrooms are delicious served with just a sprinkling of salt and pepper. Large mushroom caps may be stuffed prior to baking (see our recipe for Shrimp-Stuffed Mushrooms). Raw, sliced mushrooms also make a tasty addition to salads.

HERBED MUSHROOM SOUP

2	tablespoons butter or margarine
1	small garlic clove, minced
5 to 6	scallions, including green tops, chopped
1	cup finely chopped fresh parsley leaves
2	cups sliced fresh mushrooms (about ⅓ pound)
6	cups chicken stock or broth
1	tablespoon soy sauce
1	tablespoon lemon juice, preferably fresh
½	teaspoon salt (omit if commercial chicken broth is used)
¼	teaspoon black pepper, preferably freshly ground
1½	tablespoons cornstarch
¼	cup cold water

Melt the butter in a 3- to 4-quart saucepan or pot over medium-high heat. Add the garlic, scallions, and parsley and cook, stirring, for 3 minutes. Stir in the mushrooms and continue cooking for 3 to 4 minutes longer, or until the scallions and mushrooms are limp. Add the broth, soy sauce, lemon juice, salt, and pepper and bring the mixture to a boil. Cover the pot. Lower the heat and simmer the mixture for 25 minutes, or until the parsley is tender.

Stir together the cornstarch and water in a small cup or bowl until smooth. Stir the mixture into the soup and cook for 1½ to 2 minutes longer, or until the soup is clear and slightly thickened.

Makes 4 to 6 servings.

MUSHROOM SLAW

1¼	cups thinly sliced fresh mushrooms (about 3½ ounces)
⅓	cup mayonnaise
1	tablespoon apple cider vinegar
1	tablespoon chopped dill pickle or pickle relish
1	teaspoon celery seeds
½	teaspoon dried dillweed
¼	teaspoon salt
¼	teaspoon black pepper, preferably freshly ground
3½ to 4	cups coarsely grated or shredded green cabbage

Combine all the ingredients, *except* the cabbage, in a large bowl and stir until well mixed. Cover the bowl and refrigerate for at least 2 hours and up to 8 hours to give the flavors a chance to mingle.

A few minutes before serving time, add the cabbage to the bowl, tossing until well mixed. Let the slaw stand for 5 or 10 minutes to allow the flavors to blend.

Makes 4 to 5 servings.

CONFETTI RICE

(side dish)

  2   tablespoons butter or margarine
5 to 6 scallions, including green tops, coarsely chopped
  1   small carrot, finely chopped
  1   small celery stalk, including leaves, finely chopped
  ¼   cup finely chopped sweet green pepper
  ¼   cup finely chopped sweet red pepper (optional)
  ¾   cup coarsely chopped fresh mushrooms
  ¼   teaspoon curry powder
  ¼   teaspoon salt
  ⅛   teaspoon black pepper, preferably freshly ground
2½ cups *cooked* white or brown rice
  1   medium-sized tomato, peeled and chopped

Combine the butter, scallions, carrot, celery, green pepper, and red pepper (if used) in a large skillet over medium-high heat. Cook, stirring, for 3 minutes. Add the mushrooms and cook for 2 to 3 minutes longer, or until the vegetables are limp. Stir in the curry powder, salt, and black pepper until blended. Add the rice and cook, stirring, for 1 minute longer. Stir in the chopped tomato and continue cooking until it is just heated through.

Makes 4 to 5 servings.

SHRIMP-STUFFED MUSHROOMS

(appetizer)

This makes a delicious first course for an elegant dinner.

1½	pounds large fresh mushrooms (about 16 3-inch-diameter mushrooms)
2	tablespoons butter or margarine
4 to 5	scallions, including green tops, finely chopped
1	cup fresh bread crumbs
¾	cup (about 2 ounces) small cooked frozen shrimp, thawed and chopped
¼	teaspoon salt
⅛	teaspoon freshly ground black pepper
	Pinch of dried thyme leaves
⅓	cup light cream or half-and-half, divided
¼	cup chicken broth or bouillon (reconstituted from cubes or granules)
2	tablespoons grated Parmesan cheese

Carefully remove the mushroom stems from the caps. Arrange the caps, undersides facing up, in a lightly greased or nonstick spray-coated 11- by 7-inch baking dish. Finely chop the stems and set them aside.

Melt the butter in a medium-sized saucepan over medium-high heat. Add the scallions and cook, stirring for 3 to 4 minutes, or until they are limp. Add the chopped mushroom stems and the bread crumbs and cook, stirring frequently, for 3 minutes. Stir in the shrimp and cook, stirring, for about 2 minutes longer, or until the mushrooms are soft and the shrimp are just heated through. Stir in the salt, pepper, thyme, and 2 tablespoons of the cream until well blended. Remove the skillet from the heat and let the bread crumb-shrimp mixture cool for a minute or two. Using a teaspoon or your fingers, stuff the mushroom caps with the bread crumb-shrimp mixture, dividing it equally among them.

Combine the remaining cream with the broth and pour the mixture over the stuffed mushroom caps, being sure to moisten each one. Sprinkle the Parmesan cheese over the mushroom caps.

Cover the baking dish and bake the mushrooms in a preheated 350-degree oven for 10 minutes. Uncover the dish and continue baking for 7 to 10 minutes longer, or until the mushrooms are just tender when pierced with a fork.

Makes about 8 servings.

Okra

A relative of the ornamental hibiscus, okra is a large warm-weather plant that probably originated in the area of Africa now comprising Ethiopia. It has been enjoyed not only there, but in North Africa, the Middle East, and India for at least six or seven hundred years, and it still grows wild near the upper Nile.

African slaves very likely brought okra—or "kingombo," as those from Angola called it—to New Orleans during the late 1600s or 1700s. In time, the name was shortened to just "gombo," which helps explain why only a stew containing okra can qualify as honest-to-goodness "gumbo." Okra continues to be a staple in authentic Creole cookery, and today, nearly every family garden patch in Louisiana includes this vegetable. It is also popular in other Southern states.

Okra grows on a showy, flowering bush, and the foliage in some types can reach a height of 7 to 10 feet. The finger-like edible parts, which are sometimes called "lady fingers," are actually immature seed pods that develop from the plant's pretty red-throated yellow blooms. In most varieties, the pods are slender, deep green, and ridged and can grow up to 7 inches long (although they are usually picked for commercial use before reaching this length). One miniature type, marketed as "Chinese okra," has plumper, more rounded pods that are about 2 to 3 inches long.

Outside the South, okra is not widely eaten. Its taste is distinctive and often has

166

to be acquired. Sometimes, people also object to the slightly "slippery" texture that can develop as okra exudes its juices during cooking. This "roping" or slight thickening of the juices is simply the result of a reaction of the plant's protein to heat. Roping can be minimized by taking care when the stems are trimmed off (see Basic Preparation and Cooking for details), and by cooking the vegetable whole and only until crisp-tender. On the other hand, some cooks capitalize on the roping, relying on cut okra as a thickener for gumbos, soups, and stews.

Availability: Okra's peak season is June through August. However, it can be found in small quantities year round. Frozen okra is also available throughout the year in many supermarkets, both whole and in pieces. In addition, canned okra can sometimes be purchased, but most people find it a poor substitute for fresh or frozen.

Choosing the Best: Look for firm, rich green, unblemished okra pods, preferably less than 3 inches long. If possible, choose ones that are even shorter than this, as they will be most tender and succulent. Pods in good condition are springy and resilient when pressed, and show no signs of surface discoloration. Limpness and black or brown areas are signs of bruising or improper storage.

Nutritional Value: Okra has some vitamin A and minerals, but it is not a significant source of any essential nutrients. Being almost 90 percent water, it is very low in calories. The seeds of fully ripe pods are high in protein, but since the vegetable is normally consumed when still immature, it contributes only modest amounts of protein to the diet.

Storage: Fresh okra pods are highly perishable and should be used as soon as possible after purchase. If they must be stored, place (unwashed) in the refrigerator in a loosely closed plastic bag. The temperature should not be lower than 45 degrees or okra will darken and "burn" from the cold.

Preparation and Basic Cooking: The pods are always eaten cooked, sometimes whole and sometimes cut crosswise into short lengths. Since the edible portions form well above the ground, they usually need only a quick rinsing before use. To prepare them for cooking, simply pare away the tough stem cap at the top of each. If you wish to minimize roping, trim the caps in a cone shape and avoid cutting so deeply that the seeds inside are exposed. The pods may also be brushed with a vegetable brush to remove the very thin layer of fuzz on the skin, although most people find this step unnecessary.

The best way to keep okra from releasing its juices during cooking is to leave the pods whole. Cook them in a small amount of liquid and avoid letting them get too done. On the other hand, if you want the okra to serve as a thickener, cut up the pods so most of the juices will be released. Whole okra may be steamed or boiled (or braised) in a small amount of liquid for 5 to 8 minutes, or until crisp-tender.

Never cook okra in a copper, tin, brass, or iron pot, as these metals can react to the juices in the pods and cause them to turn dark and unappetizing looking. (The chemical reaction is completely harmless, however.)

Simple Serving Suggestions: Dress plain-cooked okra with a little butter, salt, pepper, and lemon juice. Or combine it with tomatoes in vegetable medleys and soups and along with fish or chicken in spicy gumbos and stews.

OKRA AND SWEET RED PEPPER STIR-FRY

(side dish)

The bright green of the okra pods and red of the pepper slices make this a very pretty dish. Okra fans will find it delicious and it may even win over some who think they don't care for this unusual vegetable.

2	tablespoons peanut or vegetable oil
4 to 5	scallions, white part only, cut into 1-inch lengths
½	pound trimmed fresh whole okra (preferably small pods about 2½ inches long or less)
1	large sweet red pepper, cut into 2-inch-long by ¼-inch strips
⅓	cup water
2	teaspoons soy sauce
¼	teaspoon black pepper, preferably freshly ground
1	large tomato, peeled and coarsely chopped

Heat the oil in a large skillet over high heat until very hot but not smoking. Add the scallions, okra, and sweet red pepper and cook, stirring, for 2 minutes, or until the onion is limp and the vegetables are well coated with the oil. Add the water, soy sauce, and black pepper and allow the mixture to come to a boil. Lower the heat to medium and cook the mixture, stirring occasionally, for 3 to 4 minutes longer, or until the vegetables are almost tender and most of the liquid has evaporated from the pan. Stir in the tomato and continue to cook for about 3 to 4 minutes longer, or until it has cooked down slightly and the okra and pepper strips are crisp-tender.

Makes about 4 servings.

OKRA COOKED WITH RICE AND TOMATOES

(side dish)

1 medium-sized onion, finely chopped
2 tablespoons vegetable oil
1 cup *uncooked* white rice
½ pound fresh okra, trimmed and cut crosswise into ½-inch-thick pieces (about 2 cups)
1 16-ounce can tomatoes, including juice, chopped
1¼ cups water
½ teaspoon salt
⅛ teaspoon black pepper, preferably freshly ground
4 or more drops Tabasco sauce (to taste)

In a large saucepan over medium-high heat, cook the onion in the oil, stirring, until the onion is tender but not browned. Add the rice and cook, stirring, about 1 minute longer. Then add the remaining ingredients and bring the mixture to a boil. Lower the heat, cover the saucepan, and simmer the rice mixture for about 25 minutes, or until all the liquid has been absorbed and the rice is tender. (If all the liquid has been absorbed and the rice is still not tender, add a little more water and continue cooking, covered, until the water has been absorbed.) Toss with a fork briefly before serving to evenly distribute the vegetables.

Makes about 6 servings.

OKRA SUCCOTASH

(side dish)

This colorful vegetable medley is particularly popular in some parts of the South.

2 tablespoons butter or margarine
1 small onion, finely chopped
¾ pound fresh okra, trimmed and cut crosswise into ½-inch-thick pieces (or 1 10-ounce package frozen cut okra, thawed)
1 16-ounce can tomatoes, including juice
½ cup water
1 cup frozen baby lima beans
¼ teaspoon salt
¼ teaspoon black pepper, preferably freshly ground
¾ cup loose-pack frozen corn kernels, preferably "shoepeg" or other white variety
¾ cup drained cooked or canned black-eyed peas (optional)

Melt the butter in a large saucepan over medium-high heat. Add the onion and cook, stirring, for 3 to 4 minutes, or until the onion is limp. Stir in the okra, tomatoes, water, lima beans, salt, and pepper and allow the mixture to come to a boil. Lower the heat, cover the pan, and simmer gently, stirring occasionally, for 10 to 12 minutes, or until the okra and lima beans are almost tender. Stir in the corn kernels and black-eyed peas (if used). If the mixture is very thick, also add a tablespoon or 2 more water to prevent the vegetables from sticking to the bottom of the pan. Cover the pan and simmer for 4 to 6 minutes longer, or until the corn is heated through and the okra and beans are tender.

Makes 5 to 6 servings.

INDIAN-STYLE OKRA

(side dish)

Seasoning Mixture
¼ cup water
2 tablespoons lemon juice, preferably fresh
1 tablespoon sugar
½ teaspoon ground coriander
¼ teaspoon ground cumin
⅛ teaspoon ground turmeric
⅛ teaspoon cayenne pepper
⅛ teaspoon salt

Okra and Remaining Ingredients
¾ pound fresh whole okra (or 1 10-ounce package frozen whole okra, thawed)
2 tablespoons vegetable oil
1 teaspoon mustard seeds
1 small garlic clove, minced
1 small onion, finely chopped

Combine all the seasoning mixture ingredients in a small cup or bowl. Stir until well mixed and set aside.

To prepare the okra, rinse the pods and pat them dry on paper towels. Using a sharp knife, trim off and discard the tough stem ends at the tops of the pods; work carefully and avoid cutting deeply enough to expose the seeds inside. (If frozen okra is used, drain well in a colander and pat dry on paper towels.)

Heat the vegetable oil in a large skillet over high heat until hot but not smoking. Add the mustard seeds and cook, stirring, for 1 minute. Add the garlic and onion and cook, continuing to stir, for 1 minute longer. Stir in the okra until it is coated all over with the oil. Cook, stirring, for 2 minutes. Stir in the seasoning mixture. Lower the heat to medium and continue cooking, stirring occasionally, for 6 to 8 minutes longer, or until the okra pods are crisp-tender and most of the excess liquid has evaporated from the pan.

Makes 4 to 5 servings.

TURKISH-STYLE MIXED VEGETABLES

(side dish)

Similar to the French favorite ratatouille, this dish is good both hot and chilled. Unlike its cousin, however, it features okra and Italian green beans. In fact, it's a good way to introduce these vegetables to your family if they are not familiar with them.

2 medium-sized onions, thinly sliced
2 tablespoons olive or vegetable oil
3 medium-sized zucchini, cut lengthwise in half and then into 1-inch-long pieces
2 medium-sized sweet green peppers, cut into ½-inch-wide strips
1 medium-sized eggplant, partially peeled lengthwise in stripes, and then cut into 1-inch cubes
1 9- or 10-ounce package frozen Italian green beans, slightly thawed
1 10-ounce package frozen whole okra, slightly thawed
2 16-ounce cans tomatoes, including juice, coarsely chopped
½ teaspoon salt
⅛ teaspoon black pepper, preferably freshly ground

In a large pot or a Dutch oven over medium-high heat, cook the onions in the oil, stirring, until they are tender but not browned. Add the zucchini, green peppers, and eggplant, and cook, stirring, for about 2 minutes longer. Then add the remaining ingredients and stir all the vegetables gently just until they are evenly distributed.

Continue heating until the juices come to a boil. Lower the heat, cover the saucepan, and simmer for about 25 to 30 minutes, or until all the vegetables are tender. Gently stir once or twice during the cooking period.

Makes 8 to 10 servings.

Onions

Even before the dawn of recorded history, our ancestors knew and appreciated the full-bodied flavor of the onion. A member of the lily family, this vegetable has been cultivated since man first began to sow seeds and raise plants for food.

While the onion was known to both the Babylonians and the Egyptians, the latter at first considered it a sacred plant. But by the time the great Pyramids were built, the vegetable had found its way into Egyptian cooking pots. In fact, onions were prominent in the diet of Egyptian slaves, including the Israelites.

Down through the ages, onions were thought to confer strength and health—and to ward off and cure various illnesses. Greek and Roman athletes ate them before competitions. And, in Nero's time, onion juice was rubbed on the bodies of gladiators—possibly to make the lions less anxious to take a chomp! In the Middle Ages, onion was used as a cure for dog and snake bites. In early New England, settlers hung onions on the front door to keep diseases from entering the house—much as the superstitious used garlic to ward off vampires.

Although the American Indians enjoyed a form of wild onion or garlic, the first domesticated onions were brought to the New World by the Spanish. Onion seeds also crossed the Atlantic on the Mayflower.

There are many onion varieties.

172

Chives are the mildest member of this large family. Their bulbs are not eaten, but the delicate green stalks are snipped or cut and used in salads, as a soup garnish, and in cooking.

Garlic is a small but pungent member of the onion family that has long been used in cooking. The flavor is so strong that a little goes a long way. Garlic bulbs are about the size of a golf ball and are divided into many white or purplish-white crescent-shaped "cloves." One or two of these is usually enough to flavor a whole pot of stew or soup.

Leeks, which are often unfortunately neglected by American cooks, add a distinctive flavor to soups and stews. A relatively mild onion variety, they are shaped like long, narrow cylinders with compactly rolled leaves. The base of the plant is white. The leaves become progressively darker green toward the top.

Scallions and green onions are for all practical purposes used interchangeably and are known by the former name in some parts of the country and the latter in others. Technically, however, scallions are harvested before there is any bulb formation. Green onions have developed a bit further and have small bulbs—about 1 to 2 inches across. Both have dark green cylindrical stalks, which are white at the root base.

Shallots are a small, delicately flavored member of the onion family that forms heads somewhat reminiscent of the garlic bulb. However, each head has fewer individual sections and they are more loosely joined than those of garlic. Shallots have dry, reddish skin and cloves that are tinged with purple. They are just beginning to be widely available commercially.

Storage onions are mature onions that have been dried slightly. They come in a variety of shapes and sizes—from tiny white onions (such as the pearl onion) to large Spanish, red, and Bermuda onions. In between are the medium-sized white and yellow "all-purpose" onions.

The large flat-topped Bermuda (white or yellow), Spanish (brown or yellow), and red onions are the mildest, followed by the small white onions. Globe-shaped all-purpose cooking onions are the most strongly flavored and have white, yellow, or light brown skins.

Availability: Scallions and green onions are available year round, as are garlic, leeks, and most storage onions.

Chives are not commonly found in U. S. supermarkets but are sold, growing in pots, in some specialty produce shops. They are easy to raise in the home garden and come back every spring. A small clump can be divided into several plants after a year or two. Garden chives can be harvested with shears or a knife until well into fall.

Choosing the Best: Scallions and green onions should have firm, crisp, dark green stems. Avoid bunches that are limp or have withered or brown tips.

Look for leeks that have firm white bases, tightly rolled leaves, and fresh tops. Larger leeks are not as tender as the smaller or medium-sized ones. Avoid any that have soft spots or are bruised or wilted or turning yellow at the tops.

Storage onions should be firm, with bright, smooth, dry skins. The skin should be crackly and blemish-free. Avoid onions that have begun to sprout, as well as soft, spongy onions and those with wet or soft necks or hollow woody centers. A strong oniony aroma often indicates that the vegetable has begun to deteriorate.

When shopping for garlic, look for firm, plump bulbs with solid cloves that are neither dried out nor beginning to sprout. The skin of the bulb should be dry and smooth.

Nutritional Value: Onions contain some vitamin C. They are also low in calories. Three and a half ounces of raw onion have about 38 calories. Green onions and scallions are also a good source of vitamin A.

Storage: Leeks, green onions, and scallions should be well wrapped in plastic and refrigerated. Green onions, scallions, and leeks can be stored for up to a week. Cut off leek roots and any wilted parts before storing.

Whole storage onions should be kept in a cool, dark place with good ventilation. To help prevent sprouting, they can also be wrapped in a plastic bag and refrigerated. Storage onions can be held for 3 or 4 weeks and sometimes longer.

Preparation and Basic Cooking: Leeks need to be cleaned particularly well before cooking. Because of their growth pattern, dirt and sand collect under the leaves. To clean leeks, first cut off the rootlets. Peel off 1 or 2 layers of tough outer leaves and cut away the tough upper portion of the plant (all except about an inch of the green part). Then, beginning at the green end, slice the leeks in half lengthwise (or cut down only about 2 inches). Place the leeks in a bowl of cold water and swish them around to remove dirt and sand. Change the water and repeat until there is no sand or dirt at the bottom of the bowl. Alternatively, if the leeks have not been cut all the way through, you can separate the leaf layers at the top and rinse the leeks well under cold running water. Drain the leeks before cooking.

For scallions and green onions, cut off the roots and any bruised or wilted leaf tops. Wash under cold running water and drain before slicing.

For mature onions, cut off any sprouted tops and discard. If onions make your eyes sting and tear, try holding them under running water while peeling them. (Or store onions in the refrigerator, since chilling seems to make them less likely to affect the eyes.) The flavoring substance in onions is a volatile oil that evaporates on exposure to air. Therefore, for best flavor, peel onions just before using them.

To peel small onions, cover them with boiling water and let them stand for 3 to 5 minutes. Then drain, trim off the stem and root ends, and peel. To ensure even cooking and to keep the onion intact, cut a shallow "X" into each root end. To cook whole, drop peeled onions into boiling water or bouillon. Cook for 15 to 35 minutes, depending on size. Onions are done when they are transparent and easily pierced with a fork.

Simple Serving Suggestions: Bermuda, Spanish, and red onions, scallions, green onions, and shallots are mild enough to be served raw in salads.

Scallions and green onions are also particularly good in Oriental stir-fry dishes. Medium-sized yellow onions are most often used as flavorings in soups, stews, skillet dishes, and other cooked entrées. Cooked small white onions can be served with a cream or cheese sauce or flavored with butter and minced parsley, black pepper, or dried thyme or basil leaves.

Leeks can be braised in broth and topped with some herbs and a light sprinkling of cheese.

RED ONION SOUP

Although quite different from classic French onion soup, this version is appealing in its own right. Serve it as a light main course on chilly winter days.

   3   cups vegetable or beef bouillon (reconstituted from cubes or granules) or beef stock or vegetable stock (page 86)
   3   cups water
   1½ cups dry red wine
   8   cups coarsely chopped red onions
   1   large carrot, grated or shredded
   1   garlic clove, minced
   2   bay leaves
   ½   teaspoon dried thyme leaves
   ¼   teaspoon powdered mustard
   ½   teaspoon salt
   ¼   teaspoon black pepper, preferably freshly ground
       Pinch of cayenne pepper

To Serve
       French bread, cut into ½-inch-thick slices
       About 6 tablespoons grated Parmesan cheese

Combine all the soup ingredients in a large saucepan or small Dutch oven. Cover and bring to a boil over high heat. Lower the heat and simmer the soup for about 2½ hours, stirring occasionally.

To serve, toast the slices of French bread. Ladle the soup into individual serving bowls. Float a slice of the toasted French bread on the soup in each bowl. Sprinkle each slice with about 1 tablespoon of Parmesan cheese.

Makes about 6 servings.

POTATO-LEEK SOUP

Good either hot or cold, this soup tastes best when made the day before serving. Well chilled, it makes a nice summer appetizer.

6 large leeks, white part only, well washed and thinly sliced (about 3 pounds; see Preparation directions for cleaning instructions)
1 garlic clove, minced
2 cups 1-inch peeled potato cubes
3 cups vegetable bouillon (reconstituted from cubes or granules) or vegetable stock (page 86) or chicken broth
½ teaspoon dried thyme leaves
1 teaspoon dried basil leaves
3 tablespoons butter or margarine
⅛ teaspoon powdered mustard
¾ teaspoon salt
⅛ teaspoon black pepper, preferably freshly ground
3 cups whole milk

In a large saucepan or small Dutch oven, combine the leeks, garlic, potatoes, bouillon, thyme, basil, butter, powdered mustard, salt, and pepper. Cover and bring to a boil over medium-high heat. Lower the heat and simmer, stirring occasionally, for about 30 minutes, or until the vegetables are very soft.

In a food processor or blender, purée the vegetables and stock, in batches if necessary. Return the purée to the pan. Stir in the milk and blend well. Cook over very low heat, stirring occasionally, for an additional 10 minutes. Serve hot or cold. Stir well before serving.

Makes 6 to 8 servings.

BRAISED ONIONS

(side dish)

2 tablespoons butter or margarine
1 pound small white onions, trimmed and peeled (about 20; see Preparation directions for trimming and peeling instructions)
½ cup beef broth or bouillon (reconstituted from cubes or granules) or dry red wine
1 bay leaf
   Pinch of ground thyme
⅛ teaspoon salt
⅛ teaspoon black pepper, preferably freshly ground

In a large saucepan over medium heat, melt the butter. Add the onions and cook them, stirring frequently, for about 5 minutes. Add all the remaining ingredients. Bring the liquid to a boil and cover the pan. Lower the heat and simmer the onions for 15 to 20 minutes, or until they are tender. Uncover the pan. Raise the heat and cook the sauce down until only about ¼ cup remains.

Makes 4 to 5 servings.

BROWN RICE WITH LEEKS AND CARROTS

(side dish)

This is colorful and tasty—a great way to enjoy brown rice.

1	bunch leeks (about 4 medium sized)
2	tablespoons olive or vegetable oil
1	cup *uncooked* brown rice
1	cup thinly sliced carrots
2	cups boiling water
2	tablespoons tomato paste
2	tablespoons finely chopped fresh parsley leaves
1	tablespoon lemon juice, preferably fresh
¼	teaspoon salt
⅛	teaspoon black pepper, preferably freshly ground

Clean the leeks carefully to remove any sand inside them. (To do this, cut off and discard the root and all the green part except for about 1 inch. Then, beginning at the green end, cut the leeks in half lengthwise about 2 inches down. Separate the "layers" at the top and rinse the leeks well under cold running water; then drain the leeks on paper towels.) Thinly slice the cleaned leeks.

In a large saucepan, heat the oil over medium-high heat; then cook the leeks until they are tender but not browned. Stir in the rice and sauté about 1 minute longer. Add the remaining ingredients and stir briefly. Bring the rice mixture to a boil; then lower the heat and simmer, covered, for about 40 to 45 minutes, or until all the liquid has been absorbed. Toss with a fork before serving to evenly distribute the vegetables.

Makes about 6 servings.

ONIONS IN PAPRIKA SAUCE

(side dish)

Try this zesty dish as a change of pace from creamed onions. It goes quite well with roast chicken or meat loaf.

1¼   pounds small white onions (about 24 onions)
     About ¾ cup chicken broth or bouillon (reconstituted from cubes or granules)

Paprika Sauce
   2   tablespoons butter or margarine
     Generous 2 tablespoons enriched all-purpose or unbleached white flour
   2   teaspoons paprika, preferably sweet Hungarian
     Generous ¼ teaspoon chili powder, preferably "hot"
   ⅛   teaspoon salt
   ½   cup chicken broth or bouillon (reconstituted from cubes or granules)
   ½   cup whole milk

In a large saucepan, just barely cover the whole, unpeeled onions with water and bring to a boil over high heat. Lower the heat and cover the saucepan; simmer the onions for 5 minutes. Transfer the onions into a colander and rinse them under cold water until cool enough to handle. Trim off the roots and tops and peel the onions with a sharp knife. Cut a ¼-inch-deep "X" in the root end of each onion.

Rinse out the saucepan in which the onions were simmered and return the onions to it. Add ¾ cup broth to the pan and bring the mixture to a boil over high heat. Lower the heat and gently simmer the onions, uncovered, for 3 to 6 minutes, or until they are cooked through but still slightly firm when pierced with a fork. (If you prefer softer onions, cook for 5 to 8 minutes, or until they are tender when pierced with a fork.) If necessary, add a bit more broth to the pan to prevent it from boiling dry. Transfer the onions to a serving dish with a slotted spoon.

To prepare the sauce, melt the butter in a small heavy saucepan over medium-high heat. Stir in the flour until well blended and smooth. Cook the mixture, stirring, for 2 minutes. Stir in the paprika, chili powder, and salt until smooth. Then add the broth and milk, stirring constantly until the mixture is well blended. Heat the mixture, stirring, until thickened and smooth, about 2 to 3 minutes longer. Pour the sauce over the onions and serve.

Makes 5 to 6 servings.

LEEK, BROWN RICE, AND BARLEY BAKE
WITH MUSHROOMS

(side dish)

Use this hearty dish to fill out a skimpy meal or to add interest to a fancy one. The casserole can be assembled and then baked several hours later, if desired.

1	cup *uncooked* brown rice, washed and well drained
½	cup pearl barley
3	cups beef broth or bouillon (reconstituted from cubes or granules)
1	large bay leaf
¾	teaspoon dried thyme leaves
	Generous ¼ teaspoon black pepper, preferably freshly ground
6 to 8	medium-sized leeks, including white part and 1 inch of green top
3	tablespoons butter or margarine
2	cups coarsely sliced fresh mushrooms
2	tablespoons chopped fresh parsley leaves
	Paprika for garnishing casserole top (optional)

Combine the rice, barley, broth, bay leaf, thyme, and black pepper in a large saucepan over medium-high heat. Bring the mixture to a boil. Lower the heat, cover the pan tightly, and gently simmer the mixture for 30 minutes.

Meanwhile, slice the leeks in half lengthwise. Separate the layers and rinse well under cool running water. Coarsely chop the leeks and put them in a colander. Rinse them thoroughly and let stand until well drained.

Melt the butter in a large skillet over medium-high heat. Add the leeks and sauté, stirring, for 5 to 6 minutes, or until they are tender but not brown. Add the mushrooms and parsley and cook, stirring, for 2 to 3 minutes longer, or until the mushrooms are limp. Set the mixture aside.

When the rice and barley have cooked for 30 minutes, remove the pan from the heat and fluff up the grains with a fork. Stir in the sautéed vegetables using a fork until well mixed. Turn the mixture into a lightly greased 1½-quart ovenproof casserole. Sprinkle the top lightly with paprika, if desired. The casserole may now be baked, tightly covered, in a preheated 350-degree oven for 35 to 40 minutes, or until the rice and barley are just tender. Or cover and refrigerate it for up to 8 hours before baking.

Makes 5 to 7 servings.

SCALLION AND CAULIFLOWER QUICHE

(light main dish)

- **2** cups small cauliflower flowerets
- **¾** cup finely chopped scallions, including green tops
- **⅓** cup lowfat milk
- **1** tablespoon plus 1 teaspoon cornstarch
- **½** cup lowfat small curd cottage cheese
- **½** cup part-skim ricotta cheese (if unavailable, substitute regular ricotta)
- **2** large eggs plus 1 large egg white, slightly beaten
- **⅛** teaspoon ground nutmeg
- **2 to 3** drops Tabasco sauce
- **½** teaspoon salt
- **⅛** teaspoon black pepper, preferably freshly ground
- **½** teaspoon dried basil leaves
- **¼** teaspoon powdered mustard
- **5** ounces Swiss cheese, grated (1¼ cups packed)
- **¼** cup finely chopped fresh parsley leaves
- **1** unbaked 9-inch pastry crust

Put the cauliflower flowerets in a small saucepan with ½ inch of water. Cover and bring to a boil over high heat. Lower the heat and steam the flowerets for 4 minutes. Add the scallions, cover the pan, and continue steaming the vegetables for 2 additional minutes. Drain the vegetables well in a colander.

Combine the milk and cornstarch in a medium-sized bowl. Stir to blend them well. Add the cottage cheese, ricotta, eggs and egg white, nutmeg, Tabasco sauce, salt, pepper, basil, powdered mustard, Swiss cheese, and parsley and stir to blend them well.

Arrange the cauliflower flowerets and the scallions in the bottom of the pastry shell. Pour the egg mixture over them and spread it out with the back of a large spoon. Bake the quiche in a preheated 350-degree oven for about 35 minutes, or until it is set and beginning to brown around the edges.

Makes about 6 servings.

ONION AND TUNA "QUICHE"

(main dish)

In this recipe, cooked rice is used for a low-calorie, lowfat "crust." Leeks add a very nice flavor to the filling, but if you can't find them, substitute an extra onion.

1 tablespoon butter or margarine
2 medium-sized leeks, white and very pale green part only, washed well and thinly sliced (See Preparation directions for cleaning instructions.)
1 small onion, thinly sliced
2 cups *cooked* brown or white rice, warm or cold
2 tablespoons chopped fresh parsley leaves
½ teaspoon dried thyme leaves, divided
½ teaspoon dried marjoram leaves, divided
   Pinch of salt or garlic salt
1 large egg white, lightly beaten
½ cup grated hard or semisoft cheese, such as Swiss, Jarlsburg, Muenster, mozzarella, etc., divided
1 6½-ounce can water-packed tuna, drained and coarsely flaked
3 large eggs, lightly beaten
1 cup skim, lowfat, or whole milk
⅛ teaspoon salt
   Pinch of black pepper, preferably freshly ground

In a medium-sized skillet, melt the butter over medium-high heat. Sauté the leeks and onion until they are tender but not browned; then remove them from the heat.

Meanwhile, mix together the cooked rice, parsley, *half* each of the thyme and marjoram, the pinch of salt, and the egg white. Press the rice mixture into the bottom and sides of a deep 9-inch pie pan or a 9-inch quiche dish.

Sprinkle all but 3 tablespoons of the grated cheese into the bottom of the rice shell; then evenly distribute the tuna on top. Cover with the sautéed leek mixture.

In a small bowl, mix together the 3 eggs and the milk, the remaining thyme and marjoram, and the ⅛ teaspoon salt and the pepper. Pour over the leek mixture. Sprinkle the reserved grated cheese on top.

Bake the "quiche" in a preheated 350-degree oven for about 40 to 45 minutes, or until it is set. Let the "quiche" stand about 5 minutes before serving, so that it will be easier to cut.

Makes about 4 servings.

HOT AND SPICY ORIENTAL-STYLE CHICKEN

(main dish)

Marinade and Chicken
¼ cup dry sherry
2 tablespoons soy sauce
¼ teaspoon cayenne pepper
1 teaspoon grated or minced gingerroot (or ¼ teaspoon ground ginger)
1 garlic clove, minced
1 pound boneless chicken (meat from 3 large breast halves), cut into 1-inch cubes

Vegetables
2 tablespoons vegetable oil
½ cup salted or unsalted dry roasted peanuts
1 large carrot, grated or shredded
1 large onion, finely chopped
1 large sweet green or red pepper, cut into ½-inch squares
2 celery stalks, thinly sliced

Sauce
3 tablespoons dry sherry
1 tablespoon packed light brown sugar
⅛ teaspoon cayenne pepper
1 teaspoon grated or finely chopped gingerroot (or ¼ teaspoon ground ginger)
1 tablespoon cornstarch
2 tablespoons soy sauce
   Pinch of black pepper, preferably freshly ground

To Serve
   Hot cooked white rice

In a medium-sized glass or ceramic bowl, combine all the marinade ingredients. Add the chicken and stir to coat well. Refrigerate, covered, for about 1 hour.

In a large heavy skillet over medium-high heat, heat the oil. Stir in the peanuts. Cook them in the oil, stirring, just until they begin to turn brown. Remove the peanuts with a slotted spoon and set them aside.

Remove the chicken from the marinade with a slotted spoon; reserve the marinade. Add the chicken to the oil and cook, stirring, until it becomes opaque. Add the carrot, onion, sweet pepper, and celery. Continue cooking, stirring, for about 3 minutes longer. Turn the heat to low. Cover and cook for about 5 to 7 minutes longer, stirring occasionally, or until the onion is tender.

While the chicken and vegetables are cooking, combine all the sauce ingredients in a small bowl. Add the sauce to the pan along with any remaining marinade and the reserved peanuts. Raise the heat to medium-high and stir until the sauce thickens. Serve immediately over hot cooked white rice.

Makes 4 to 5 servings.

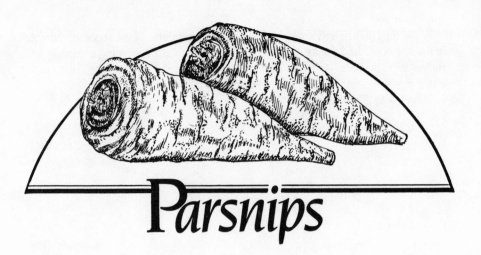

Parsnips

Down through the ages, kings and commoners alike have dined on parsnips. In ancient Rome, they were reserved for the aristocracy, who liked them drowned in honey or combined with fruit in little cakes. The Roman emperor Tiberius was so fond of their sweet, nut-like flavor that he had them specially imported from Germany when they were out of season in Italy.

In the Middle Ages, parsnips were a dietary staple of the common people and were often cooked with such thrifty ingredients as salted cod and pickled eel.

By the Renaissance, however, the upper classes had again discovered this root vegetable. Until displaced in popularity by the potato, parsnips were *de rigueur* as an accompaniment to such banquet dishes as boiled swans and roasted side of beef.

A member of the parsley family—which also includes celery, carrots, parsley, and fennel—the parsnip is native to Eastern Europe and requires a long, cool growing season. The edible part of the plant is a large taproot, which resembles a whitish carrot. Generally, parsnips are not harvested until after the first frost and are not harmed if left in the ground during the winter. In fact, the vegetable's characteristically sweet taste does not develop until after cold weather sets in and stimulates the root to convert its starch to sugar.

Parsnips were brought to the New World by the earliest colonists, who used them in a wide variety of dishes and even fermented them to make wine. Today, they are much less popular here, but are still grown commercially in the northern half of the United States.

Availability: Fresh parsnips are available all year round with the peak supply in late winter and early spring.

Choosing the Best: Choose small to medium-sized parsnips, as larger ones are likely to have woody cores. Look for firm, well-shaped parsnips. Avoid any that are discolored, shriveled, or bruised. Badly wilted, flabby parsnips are likely to toughen during cooking.

Nutritional Value: Parsnips are not a good source of vitamins and minerals but are low in calories. Three and a half ounces of cooked parsnips have 66 calories. Parsnips should not be eaten in large quantities because they contain carcinogenic chemicals called coumarins.

Storage: Parsnips keep well in the refrigerator, unwashed in a plastic bag, for 1 to 3 weeks.

Preparation and Basic Cooking: Wash the parsnips and trim off the stem ends. Parsnips can be peeled or scraped before cooking, or boiled whole and unpeeled. The skin is then easy to strip off after cooking.

Depending on how they will be used, parsnips can be sliced, cut into strips, or diced. Pieces should be of uniform size for even cooking. Whole parsnips should be boiled for 20 to 25 minutes, or until they are tender. Pieces will take 5 to 10 minutes.

Serving Suggestions: Parsnips add flavor to soups and stews. They can also be puréed and served like mashed potatoes. Seasonings that go well with parsnips include ginger, tarragon, thyme, parsley, dill, cinnamon, ginger, and lemon butter.

GLAZED PARSNIPS

(side dish)

1   pound parsnips, trimmed, peeled, and cut into ¼-inch-thick diagonal slices
1¾   cups beef broth, preferably homemade, or bouillon (reconstituted from cubes or granules)
⅛   teaspoon salt
1   tablespoon butter or margarine
2   teaspoons sugar
2   teaspoons enriched all-purpose or unbleached white flour
2   tablespoons light cream or half-and-half

In a heavy medium-sized saucepan over high heat, combine the parsnips, broth, and salt. Bring to a boil. Then lower the heat and simmer the parsnips, uncovered, for 6 to 9 minutes, or until they are just barely cooked through. Immediately remove the pan from the heat. Transfer the parsnips to a serving bowl with a slotted spoon. Reserve the remaining pot liquid in a cup.

In the same saucepan, melt the butter over medium-high heat. When the butter is hot but not smoking, add the sugar, stirring. Cook, stirring constantly, for about 1 minute longer, or until the sugar turns light golden brown. Watch carefully for scorching and lift the pan from the heat if necessary.

Remove the pan from the heat and stir in the flour. Return the mixture to the heat and cook, stirring vigorously, for 30 seconds longer. Gradually add the reserved pan liquid, and then the cream, stirring until the sauce is smooth and well blended. Return the parsnips to the saucepan and cook for 1 or 2 minutes longer, or until they are piping hot. Transfer the parsnips and sauce to a serving bowl and serve.

Makes 4 to 5 servings.

SHERRIED PARSNIPS AND CARROTS

(side dish)

1 tablespoon butter or margarine
⅔ to ¾ cup (approximately) chicken broth or bouillon (reconstituted from cubes or granules)
¼ cup orange juice
1 teaspoon packed light or dark brown sugar
⅛ teaspoon ground ginger
Pinch of ground mace
⅛ teaspoon salt
¾ pound carrots, cut into 2-inch-long by ¼-inch-thick sticks
¾ pound parsnips, cut into 2-inch-long by ¼-inch-thick sticks
2 tablespoons sweet sherry

Combine the butter, ⅔ cup broth, orange juice, sugar, spices, and salt in a large skillet over medium-high heat. Heat the mixture, stirring, until the sugar dissolves. Add the carrots and bring the mixture to a boil. Lower the heat and simmer, stirring occasionally, for 8 minutes. Add the parsnips and continue cooking for 10 to 13 minutes longer, or until the vegetables are almost tender and most of the liquid has evaporated; if necessary, add a bit more broth to prevent the pan from boiling dry. Add the sherry and cook for 2 to 3 minutes longer, or until most of the liquid has evaporated and the vegetables are just tender. Serve the vegetables along with any remaining pan liquid.

Makes 5 to 6 servings.

Peas

Peas have been human food since prehistoric times; however, until relatively recently, they were usually eaten only after they had been dried and cooked into soup, porridge, and the like.

When Catherine de Medici married King Henry II of France, she introduced his countrymen to small, sweet, fresh *piselli novelli* ("new peas"), which she had brought from her home in Florence, Italy. During the reign of Louis XIV, these gems were enthusiastically adopted by all the fashionable French, who dubbed them *petits pois*, the name still in use worldwide to describe a very tasty type of baby pea.

It is said that Christopher Columbus carried the first peas to the New World. The native Indians eventually spread cultivation of the pea throughout North America. Later, Thomas Jefferson had at least thirty different varieties at Monticello.

The moist, cool climate of England is ideal for growing fresh peas; thus, over the years, many new strains have been developed there. For this reason, ordinary garden peas are often called "English peas" to differentiate them from other types, such as dried peas. In the American South, in particular, the term "English peas" or "green peas" is used, because "pea" alone usually means cowpea (such as the black-eyed pea).

Most fresh English peas are picked when they are still immature and their sugar

188

content is highest. As they continue to ripen, some of the sugar turns to starch, and the amount of protein increases. Peas intended for drying are left to mature.

English peas must be shelled before eating. However, two other types of peas that can be consumed pod and all have recently become popular in many parts of our country. Descriptions of these types follow:

Snow peas contain very small, underdeveloped peas inside shiny, flat, tender pods. Because Americans usually associate snow peas with Oriental cuisine, they are frequently called "Chinese pea pods." However, they are also a specialty of the Pennsylvania Dutch, who call them Mennonite Peas. And they are quite popular in Europe, where they are known by their descriptive French name, *Mange-Tout* ("eat all"). They are sometimes called "sugar peas."

Sugar snap peas are a cross between English garden peas and snow peas, with the best properties of both. They have tender, edible pods that contain plump, rounded peas. Sugar snaps have a very sweet flavor, and can be cooked whole or sliced.

Availability: Fresh English (garden) peas are seen only in limited quantities, from late March to May. In some areas, locally grown varieties may be around from August through October. Snow peas are available year round in some areas, except possibly in midwinter. They peak from May through September. Sugar snap peas peak from late March through June, though they are not yet widely distributed.

Frozen English peas, *petits pois,* and snow peas can all be purchased throughout the year. The first two are quite good; in fact, peas frozen at their peak may even be better than over-the-hill fresh ones. Frozen snow peas, however, lose one of their most appealing properties, the characteristic crunch of the pods. Canned green peas are also available year round; however, most people find them to be far inferior to the frozen ones.

Choosing the Best: When choosing fresh English peas, look for well-filled pods with even, dark green coloring. If the peas are young and not too mature, two pods rubbed together should squeak. Open one pod and check the peas within; they should be shiny, evenly sized, and fresh smelling. Avoid yellowing, overmature pods as they probably contain tough, starchy peas. Also pass up those with any sign of decay such as mold.

Both snow peas and sugar snap peas should be an even, bright green color. They should also be firm enough to snap easily when bent in half. Snow peas should have flat pods with barely developed peas within. Avoid those that have highly visible bulges from peas, as they are overmature and will be tough. Sugar snap pea pods, on the other hand, should be plump with developed peas.

Nutritional Value: English peas are a good or fair source of many nutrients, including protein, iron, potassium, calcium, phosphorus, thiamine, riboflavin, niacin, and folic acid, as well as vitamins A, C, and E. Edible-podded peas have less of most nutrients, but are also lower in calories. All peas are high in fiber.

Storage: After picking, the sugar in fresh peas rapidly turns to starch, ruining the subtly sweet flavor. So it's best to eat them as soon as possible. Also, the pods will begin to decay. Store fresh peas in a cool, moist place, such as a refrigerator crisper, and use them within 2 or 3 days. Do not shell English peas until just before cooking.

Preparation and Basic Cooking: Shell fresh English peas by breaking the stem end and pulling it down one side of the pod; then slide your finger inside the pod to push out the peas. Boil or steam them in a small amount of water for about 4 to 10 minutes, or just until they are tender. The time will vary quite a bit depending on the size and age of the peas.

For snow peas and sugar snap peas, break the stem end of the pod; then pull it down to remove the string running along the side. Sugar snap peas may need stringing on both sides of each pod; whereas some very young snow peas may not need it at all.

Snow peas and sugar snap peas are good to eat raw, or they can be steamed for about 3 to 5 minutes, or just until they are very brightly colored and crisp-tender. They taste best if not cooked until limp. Both of the edible-pod types are also very good when stir-fried. In a wok or a large skillet, heat a few tablespoons of oil and a pinch of salt until very hot. Add the trimmed, whole pods, and stir-fry just until they are crisp-tender, about 2 minutes.

Simple Serving Suggestions: Hot, cooked, young English peas are delicious served with butter or hollandaise, and they go well with sautéed tiny onions or sliced mushrooms. Cold peas are perfect in rice salads and other vegetable salads.

Raw snow peas can be opened on one side and stuffed with an herbed ricotta cheese mixture for a creative hors d'oeuvre. Raw snow peas and raw sugar snap peas can be served as snacks with dips. Both are naturals in stir-fried dinners with other vegetables, and perhaps meat or poultry.

RICE SALAD WITH PEAS

2 tablespoons vegetable oil
2 tablespoons apple cider vinegar
⅛ teaspoon powdered mustard
¼ teaspoon salt
Pinch of black pepper, preferably freshly ground
3 cups hot *cooked* long-grain rice (1 cup uncooked)
3 tablespoons mayonnaise
3 tablespoons instant nonfat dry milk powder
Generous ⅓ cup commercial buttermilk
2 tablespoons Dijon-style mustard
½ teaspoon sugar
1 celery stalk, finely chopped
1½ cups loose-pack frozen peas, cooked according to package directions
¾ cup seeded and coarsely chopped cucumber
¼ cup finely chopped red onion
1 tablespoon finely chopped red radish
Additional sliced radishes for garnish (optional)

In a large bowl, combine the oil, vinegar, powdered mustard, salt, and pepper and mix well. Stir in the cooked rice. Allow the rice to marinate at room temperature for 30 minutes.

In a small deep bowl, mix together the mayonnaise, milk powder, buttermilk, Dijon mustard, and sugar. Blend well, using a wire whisk, if necessary. Stir the dressing into the rice. Add all the remaining ingredients, *except* the optional radish slices, and toss gently with a fork until they are well combined. Refrigerate for at least 2 hours before serving. If desired, garnish with radish slices.

Makes 7 to 8 servings.

COUNTRY-STYLE PEAS AND CARROTS

(side dish)

Cabbage adds flavor and nutritional value to this easy dish.

2 tablespoons butter or margarine
4 cups shredded green or Savoy cabbage (about 1 pound)
¼ teaspoon dried thyme leaves
¼ teaspoon dried tarragon leaves
½ cup chicken broth or bouillon (reconstituted from cubes or granules) or vegetable stock (page 86)
3 cups loose-pack mixed frozen peas and diced carrots
¼ teaspoon salt
   Pinch of black pepper, preferably freshly ground

In a large saucepan over medium-high heat, melt the butter. Add the cabbage, thyme, and tarragon and cook, stirring often, for about 5 minutes, or until the cabbage is soft and wilted. Add the remaining ingredients and bring the mixture to a boil. Lower the heat, cover the saucepan tightly, and steam the vegetables for about 8 minutes, or until they all are heated through and tender. Stir once or twice during the cooking.
   Makes about 6 servings.

MIXED ORIENTAL VEGETABLE STIR-FRY

(side dish)

Sauce
   2 tablespoons cold water
   1½ tablespoons soy sauce
   ¾ teaspoon cornstarch

Vegetables
   2 cups very coarsely chopped celery-cabbage
   2 cups medium-sized broccoli flowerets
   2 cups medium-sized cauliflower flowerets
   2 medium-sized celery stalks, cut into ¼-inch-thick diagonal slices
   1 small sweet red pepper, cut into ½-inch pieces (if unavailable, substitute 1 small sweet green pepper)
   1½ cups fresh snow peas, stem ends removed (or substitute 1 6-ounce package frozen snow peas)

4 to 5   scallions, including green tops, coarsely chopped
    2   tablespoons peanut or vegetable oil
    1   large garlic clove, minced
    1   cup fresh bean sprouts

Stir together the water, soy sauce, and cornstarch in a small bowl or cup. Set aside.

Combine the celery-cabbage, broccoli, cauliflower, celery, pepper, snow peas, and scallions in a large bowl. Set aside.

In a 12-inch diameter (or larger) skillet or sauté pan, heat the oil over high heat until hot but not smoking. Add the garlic and cook, stirring, for 30 seconds. Stir in the vegetables reserved in the bowl. Cook, stirring constantly, for 3 minutes. Add the bean sprouts and cook for 30 seconds longer. Briefly stir the sauce and add it to the pan. Cook, stirring, until the mixture comes to a boil, thickens slightly, and is clear. Transfer the vegetables and sauce to a serving bowl.

Makes 4 to 6 servings.

INDIAN-STYLE PEAS AND POTATOES

(side dish)

Reminiscent of an Indian dish called Allo Mattar, this interesting combination of peas and potatoes is hearty enough for a light main course.

    2   tablespoons vegetable oil
    1   medium-sized onion, coarsely chopped
    1   large garlic clove, minced
1½   cups beef or vegetable bouillon (reconstituted from cubes or granules) or vegetable stock (page 86)
    2   teaspoons finely chopped fresh mint leaves (or 1 teaspoon dried mint leaves)
    1   teaspoon ground turmeric
    ½   teaspoon ground ginger
    ½   teaspoon salt
    ¼   teaspoon crushed hot red pepper (or more to taste)
    4   cups ½-inch raw peeled potato cubes (about 1½ pounds)
    2   cups loose-pack frozen peas

In a large heavy skillet over medium-high heat, combine the oil, onion, and garlic. Cook, stirring, until the onion is tender. Add the bouillon, mint, turmeric, ginger, salt, and hot pepper flakes. Stir to combine well. Cover, lower the heat, and simmer gently for about 10 minutes. Add the potatoes and stir to coat them with the bouillon mixture. Simmer the potatoes, covered, for about 20 minutes, or until they are almost tender. Stir occasionally and check the liquid. If the mixture seems too dry, add a bit more water. Stir in the peas. Simmer for an additional 7 to 10 minutes, or until the potatoes are tender. Stir the mixture occasionally. If the mixture seems to be boiling dry, add a little more water.

Makes 4 to 6 servings.

CHICKEN POT PIE

(main dish)

Whole wheat flour and sesame seeds add flavor to the crust in this favorite American recipe.

Sesame Pastry Crust
- ¾ cup enriched all-purpose or unbleached white flour
- ½ cup whole wheat flour
- 1 tablespoon sesame seeds
- ¼ teaspoon salt
- ⅓ cup (5⅓ tablespoons) butter or margarine
- 3 to 4 tablespoons cold water (or more if needed)

Filling
- 4 small unpeeled "new"' red potatoes, scrubbed, cut into ½-inch cubes (about 2½ cups), steamed or boiled until almost tender, and well drained
- 2 cups loose-pack mixed frozen peas and diced carrots, slightly thawed
  About 2 to 2½ cups diced cooked chicken (to taste)
- 2½ tablespoons butter or margarine
- 1 small onion, finely chopped
- ¼ cup enriched all-purpose or unbleached white flour
- 1¾ cups chicken broth or bouillon (reconstituted from cubes or granules)
- ¼ teaspoon dried basil leaves
- ⅛ teaspoon dried thyme leaves
- ¼ teaspoon salt (optional, or to taste)
  Pinch of black pepper, preferably freshly ground

Glaze (optional)
- 1 egg white, lightly beaten

Grease well or coat with nonstick spray an 8- to 9-inch-square or 10-inch-round baking pan (or baking dish) that is 1½ to 2 inches deep. Set aside.

To prepare the pastry for the top of the pie, combine the flours, sesame seeds, and salt in a medium-sized bowl. Cut in the butter with a pastry blender, two knives, or your fingertips until the mixture resembles coarse crumbs. Sprinkle it with 3 tablespoons of the water while mixing with a fork. If the pastry seems to be very dry, add more water as needed. Gather the pastry into a ball. Roll it out between 2 sheets of wax paper to a 9- to 10-inch square or an 11-inch circle (depending on the size and shape of the pan). Refrigerate the pastry in the wax paper while preparing the filling.

Put the potatoes, peas and carrots, and chicken in the prepared pan, and mix gently. Set aside.

Melt the butter in a medium-sized saucepan over medium-high heat. Add the onion and cook, stirring, until it is tender but not browned. Add the ¼ cup flour and cook, stirring constantly, for 1 minute. Vigorously stir in the broth and cook while stirring, until the sauce thickens and boils. Stir in the herbs and seasonings. Pour the sauce over the chicken and vegetables in the dish and stir so it is evenly distributed.

Remove the top sheet of wax paper from the chilled pastry and invert the pastry over the pan. Peel off the second sheet of wax paper. Crimp the pastry to the edge of the pan. If desired, brush the top of the pastry with the beaten egg white for a glaze. Use a small, sharp knife to cut several small slits in the pastry for steam to escape. Bake the casserole in a preheated 425-degree oven for 20 to 25 minutes, or until the pastry is crisp.

Makes about 4 servings.

Peppers

When Columbus set off on his first journey to the New World, he was expecting to arrive in India, where he could gather up a supply of the expensive spice called black pepper. When he reached the West Indies, however, what he found instead was a fiery chili that the natives used in cooking. Apparently, Columbus didn't want to come home empty handed, so he filled the hold of one of his ships with the new discovery. Back in Europe, his exotic cargo quickly became known as "Spanish Pepper," although it is completely unrelated botanically to the Indian condiment he sought.

While the pepper, or Capsicum, family is quite large and varied, there are two main types. One is sweet and the other hot. Both were cultivated by the native Americans for more than 2,000 years before Columbus arrived. Although eaten as a vegetable, pepper is technically a fruit.

Sweet peppers are often referred to as "bell peppers" because of their characteristic shape. They are green, yellow-green, or even bright yellow at maturity. Most varieties turn red as they ripen further. Bell peppers are eaten at both the mature and ripe stages. Those that turn red become sweeter as they change color.

The hotter types are called chili or cayenne peppers. Usually smaller and thinner skinned than the sweet ones, they come in a variety of shapes. Some are long and slender. Others grow almost as large as the sweet varieties. Hot peppers can be purchased green, yellow-green, or red, but are most often used after they have turned red. (Many are also sold dried.)

Availability: Peppers are available all year round, but are most abundant in the markets from June through October. Ripe red bell peppers are often sold only in summer and early fall. Yellow bell peppers are sold in some areas.

Choosing the Best: Select peppers that are firm, glossy, well shaped, and thick skinned for their variety. Pale color is a sign of immaturity and should be avoided. Also eschew blemished, shriveled, or limp ones. Quality is unaffected by size. However, the larger the pepper, the less waste proportionally.

Nutritional Value: Both hot and sweet peppers are a good source of vitamins C and A, although red peppers are higher in vitamin A than green ones.

Storage: Peppers should be used fairly quickly after they are purchased. Green peppers can be held in the refrigerator for up to a week. Red peppers should be used within 3 days.

Basic Preparation: For sweet peppers, rinse and remove the stem, seeds, and pith. For hot peppers, rinse and then follow the recipe directions. Use rubber gloves when handling hot peppers as their juice can burn the skin.

Simple Serving Suggestions: Sweet peppers are rarely served alone but are a welcome ingredient in salads and many cooked dishes. They can be used whole, cut in half, diced, or cut into strips or rings as desired. Often they are stuffed with meat and vegetable combinations and baked. The peppers can also be stuffed with a mixture of meat and rice and braised in beef broth on top of the stove. These are delicious hot or cold.

ROASTED GREEN AND RED PEPPER SALAD

Chilled, cooked vegetable salads, such as this, are very popular in the Middle East. Often, several different types are served at one meal. For the following easy salad, sweet peppers are roasted using a broiler or gas flame, and then seasoned with a vinaigrette-type dressing.

(Note: Unseasoned peppers that have been roasted, peeled, and cut into strips may be frozen for later use—a good way to store away some sweet red peppers during their short season.)

3	medium-sized sweet green peppers
3	medium-sized sweet red peppers
1 to 2	garlic cloves, very finely minced or pressed
3	tablespoons finely chopped fresh parsley leaves
3	tablespoons olive oil
1½	tablespoons lemon juice, preferably fresh
	Generous ¼ teaspoon salt
	Freshly ground black pepper to taste
	Pinch each of ground cumin and paprika (optional)

Put the peppers in a foil-lined baking pan (such as a jelly roll pan), and place the pan in a preheated broiler about 6 inches from the heating element. Broil the peppers, rotating them often with tongs, until the skins are completely blistered and charred, about 15 to 25 minutes. The red peppers may take a little less time than the green ones, so watch them carefully. (Alternatively, each pepper may be individually speared with a fork and roasted by rotating it over a gas flame until it is blistered and charred, about 3 to 4 minutes.)

Use tongs or a large spoon to transfer the roasted peppers to a brown paper bag. Fold over the top of the bag, and let the peppers cool. The steam in the bag helps loosen the skins. (Moisture may seep through the bottom of the bag, so keep it in the sink if possible.) Rinse each pepper under cool running water, while you remove and discard the skin, stem, and seeds. (Be careful of any hot steam that may remain inside the peppers.) Don't worry if the tender pepper flesh breaks into pieces. Drain the pepper flesh well (in a colander if desired).

Cut the pepper flesh into approximately ½- by 1½-inch strips. Put the strips in a bowl with the remaining ingredients and toss to combine. Cover and chill the salad for several hours or overnight to give the flavors a chance to blend. Toss again before serving.

Makes about 6 servings.

PEPPER AND ONION SAUTÉ

(side dish or garnish)

1   large sweet red pepper, cut into ½-inch squares (if unavailable, substi-
      tute 1 large sweet green pepper)
1   large sweet green pepper, cut into ½-inch squares
1   large onion, finely chopped
1   garlic clove, minced
2   tablespoons olive oil
¼   teaspoon salt
⅛   teaspoon black pepper, preferably freshly ground
¼   teaspoon dried oregano leaves (optional)
¼   teaspoon dried basil leaves (optional)

In a large skillet over medium-high heat, combine the red and green peppers, onion, and garlic with the oil. Cook, stirring, until the onion is soft, about 5 minutes.
Add the salt and black pepper, along with the herbs, if desired. Continue to cook,
stirring, until the onion begins to brown, about 5 minutes. Serve as a side dish or as
a garnish for hamburgers or other hot sandwiches.

Makes about 4 servings.

PASTA WITH PEPPERS AND ONIONS

(side dish)

*This simple dish, made with spaghetti, is a nice accompaniment to our Italian Veal
Skillet and other Italian main dishes.*

1   pound thin spaghetti
1   medium-sized sweet green pepper, cut into ½-inch squares
1   large onion, finely chopped
1   garlic clove, minced
½   teaspoon salt
     Generous ⅛ teaspoon black pepper, preferably freshly ground
1   14- to 16-ounce can tomatoes (preferably Italian plum tomatoes), includ-
      ing juice, chopped
½   cup grated Parmesan cheese

In a large pot, bring 4 quarts of water to a boil. Add the pasta, green pepper, onion,
and garlic. Boil, uncovered, until the pasta is tender, about 9 to 11 minutes. Drain
the pasta and vegetables well in a colander. Return them to the pot and stir in the

salt, black pepper, and tomatoes. Cook, uncovered, over medium heat, stirring, until the tomatoes are hot, about 1½ to 2 minutes. Transfer the pasta mixture to a serving dish and toss with the cheese.

Makes 4 to 6 servings.

CHINESE-STYLE PEPPER STEAK

(main dish)

Marinade
2   tablespoons dry sherry
2   tablespoons soy sauce
1   tablespoon water
1   teaspoon peanut or vegetable oil
¼   teaspoon ground ginger
¼   teaspoon sugar

Meat, Vegetables, and Sauce
1   pound lean steak, such as flank steak, about ½ to ¾ inch thick, trimmed of all fat
½   cup beef broth or bouillon (reconstituted from cubes or granules)
1   tablespoon cornstarch
1 to 2   tablespoons peanut or vegetable oil
1   medium-sized onion, cut in half lengthwise and then into thick wedges and separated into pieces
2   medium-sized sweet green peppers (or use 1 sweet green pepper and 1 sweet red pepper), cut into 1-inch squares
2   scallions, including green tops, thinly sliced on the diagonal

To Serve
   Hot cooked white or brown rice

Mix together all the marinade ingredients in a medium-sized bowl. Cut the steak crosswise into 2-inch pieces; then cut it with the grain into ⅛- to ¼-inch-thick slices. (This is easiest to do if the meat is slightly frozen.) Toss the meat slices with the marinade, and marinate the meat for at least 15 minutes and up to several hours (refrigerate if longer than 30 minutes), stirring occasionally.

Shortly before serving time, mix together the broth and cornstarch and set aside. Heat 1 tablespoon of the oil in a large skillet or wok over medium-high to high heat. Stir-fry the onion until it is almost tender. Add the green peppers and scallions, and continue stir-frying about 1 minute, or until the green pepper is almost crisptender. Use a slotted spoon to temporarily transfer the vegetables to a bowl.

If a skillet or wok without nonstick coating is used, heat the remaining 1 tablespoon of oil. (The oil is not necessary with nonstick cookware.) Add the meat and its marinade to the pan, and stir-fry until the meat is just very slightly pink, about 2 to 3 minutes. Return the vegetables to the pan and stir-fry for about 1 minute, thoroughly mixing the meat and vegetables. Add the reserved bouillon mixture and stir until the sauce thickens and boils. Serve immediately with the rice.

Makes about 4 servings.

Variation

CHINESE-STYLE CHICKEN AND PEPPERS

The classic recipe above is just as tasty when chicken is substituted for the steak. Use 1 pound boned and skinned chicken breast meat (from about 4 medium-sized breast halves), and cut it into 1-inch squares. Marinate the chicken as directed above for the steak. When stir-frying, cook the chicken until it is opaque and firm. The marinade will tint the chicken brown.

Makes about 4 servings.

MEXICAN-STYLE STUFFED PEPPERS

(main dish)

4	large sweet green peppers, cut in half lengthwise and seeded
1	medium-sized onion, finely chopped
1	garlic clove, minced
1	pound lean ground beef
1	15-ounce can tomato sauce
1	15- to 16-ounce can red kidney beans, well drained
1½	cups frozen loose-pack corn kernels
1	teaspoon chili powder (or to taste)
⅛	teaspoon black pepper, preferably freshly ground

Bring a large pot of water to a boil over high heat. Add the peppers, lower the heat, and boil, covered, for 2 to 4 minutes, depending on the crispness desired. Drain the peppers and set them aside.

Meanwhile, in a large skillet over medium heat, combine the onion, garlic, and ground beef. Cook, stirring and breaking up the meat with a spoon, until the meat is brown. Drain off and discard any excess fat.

Add all remaining ingredients, *except* the peppers. Stir to mix well. Simmer the meat mixture, covered, for 10 to 15 minutes. Arrange the drained green peppers in a shallow baking dish. Spoon some meat and corn mixture into each pepper, dividing it evenly. Bake, uncovered, in a preheated 350-degree oven for about 25 minutes, or until the meat mixture is heated through and bubbly.

Makes 4 servings.

TANGY HAMBURGER SKILLET

(main dish)

1	pound lean ground beef
1	large onion, finely chopped
1	garlic clove, minced
1	large sweet green pepper, coarsely chopped
1	celery stalk, thinly sliced
3	8-ounce cans tomato sauce
1	cup frozen French-style green beans
¼	cup finely chopped fresh parsley leaves
1	teaspoon prepared mustard
1	bay leaf
½	teaspoon dried marjoram leaves
	Scant ½ teaspoon dried thyme leaves
2 to 3	drops Tabasco sauce
½	teaspoon salt
¼	teaspoon black pepper, preferably freshly ground

To Serve
   Hot cooked white or brown rice

In a large heavy skillet, combine the ground beef, onion, garlic, green pepper, and celery. Cook, stirring, over high heat, breaking up the meat with a spoon, until the meat is browned and the onion is tender. Spoon off and discard any excess fat. Add all the remaining ingredients, *except* the rice. Stir to mix well. Cover, lower the heat, and simmer for 25 to 30 minutes, or until the vegetables are tender and the flavors are blended. Serve over hot cooked rice.
   Makes 5 to 6 servings.

CHICKEN AND CHILIES RELLENO CASSEROLE

(main dish)

Filling

2 tablespoons vegetable oil
¾ pound chicken breast meat, cut in ½-inch cubes (about 3 small breast halves)
2 garlic cloves, minced
1 medium-sized onion, coarsely chopped
2 4-ounce cans (net weight) chopped green chili peppers, undrained
1 15-ounce can tomato sauce
1 15- to 16-ounce can red kidney beans, well drained
⅛ teaspoon black pepper, preferably freshly ground
1 teaspoon chili powder
 Pinch to ⅛ teaspoon cayenne pepper (optional)
4 ounces mild Cheddar or Monterey Jack cheese, grated (1 cup packed)

Cornmeal Topping

1 cup yellow cornmeal
¼ cup enriched all-purpose or unbleached white flour
1 teaspoon baking powder
½ teaspoon baking soda
½ teaspoon chili powder
¼ teaspoon salt
1 cup commercial buttermilk
2 tablespoons vegetable oil
1 large egg, lightly beaten

For the filling, heat the oil in a large skillet over medium heat. Add the chicken, garlic, and onion. Cook, stirring, until the meat is cooked through and the onion is tender, about 15 minutes. Turn the heat to low and add all the remaining filling ingredients, *except* the cheese. Continue to simmer, uncovered, very gently while preparing the cornmeal topping. Stir occasionally.

For the cornmeal topping, combine the cornmeal, flour, baking powder, baking soda, chili powder, and salt in a medium-sized bowl. In a small bowl or cup, combine the buttermilk, oil, and egg. Add this to the cornmeal mixture. Stir until the dry ingredients are just moistened.

To assemble the casserole, spread the filling mixture in a 12- by 7-inch (or similar) glass baking dish. Sprinkle with the cheese. Spoon the topping mixture over the chicken mixture.

Bake in a preheated 425-degree oven for 16 to 20 minutes, or until the topping is golden brown.

Makes 6 to 8 servings.

"OVEN-FRIED" FISH WITH SALSA

(main dish)

Just as crispy and delicious as pan-fried fish, this dish is easier to prepare and much lower in fat. The Mexican-style "salsa" (sauce) adds color and flavor.

Fish

1	pound fresh or frozen (thawed) thin, skinless sole or flounder fillets
	Approximately 3 tablespoons enriched all-purpose or unbleached white flour
1	large egg
2	teaspoons water
	Approximately ½ cup yellow cornmeal

Salsa

1	tablespoon vegetable oil
1	medium-sized onion, finely chopped
1 to 2	garlic cloves, minced
1	medium-sized sweet green pepper, diced
1	16-ounce can tomatoes, including juice, chopped
1	teaspoon chili powder (or to taste)
¼	teaspoon ground cumin (optional)
⅛	teaspoon salt

Topping

2	ounces Colby, Monterey Jack, or mild Cheddar cheese, grated (½ cup packed)

Rinse the fish well. If the fillets are large, split each one lengthwise along the midline. Lightly coat each fillet with flour. Beat the egg with the water; then dip each fillet into the mixture, letting the excess drip off. Next, coat each fillet with cornmeal. Let the coated fillets dry on wax paper while the *salsa* (sauce) is prepared.

For the *salsa*, heat the oil in a medium-sized saucepan over medium-high heat; then cook the onion and garlic until they are tender but not browned. Add the green pepper and cook about 1 minute longer. Then stir in the tomatoes and their juice, chili powder, cumin (if used), and salt. Simmer, stirring occasionally, for about 5 to 10 minutes, or until the sauce is slightly thickened.

Meanwhile, preheat the broiler. Lay out the prepared fish fillets slightly separated on a lightly oiled baking sheet. Then turn each fillet over so that both sides are lightly coated with a bit of oil. Broil the fillets 4 to 6 inches from the heating element for about 3 to 5 minutes on each side, or until they are lightly browned and crusty.

Transfer the fillets to a heatproof serving dish. Spoon the *salsa* over the fillets. Sprinkle the cheese over the *salsa*. Place the dish under the heated broiler element just until the cheese melts.

Makes about 4 servings.

CHICKEN CREOLE

(main dish)

1   pound chicken breast meat, cut into 1-inch cubes (meat from 3 large breast halves)
1   teaspoon dried basil leaves
½   teaspoon dried thyme leaves
     Generous ¼ teaspoon salt
⅛   teaspoon cayenne pepper (or to taste)
2   tablespoons butter or margarine
1   medium-sized onion, finely chopped
1   medium-sized sweet green pepper, cut into ¾-inch squares
2   celery stalks, thinly sliced
1   garlic clove, minced
1   15-ounce can tomato sauce
¾   cup canned tomatoes, including juice, chopped
¼   cup finely chopped fresh parsley leaves
2   tablespoons dry white wine
2   bay leaves
1   teaspoon dried marjoram leaves
     Pinch of black pepper, preferably freshly ground

To Serve
   Hot cooked white or brown rice

In a small bowl, toss the chicken cubes with the basil, thyme, salt, and cayenne and set aside.

In a large skillet over medium heat, melt the butter. Cook the onion, green pepper, celery, and garlic until the onion is tender. Push the vegetables to one side and cook the chicken in the butter until the chicken is opaque. Add all the remaining ingredients, *except* the rice, to the skillet. Bring to a boil. Then cover the skillet and lower the heat. Simmer for about 35 minutes, or until the flavors are well combined and the celery is tender. Serve over hot cooked rice.

Makes about 5 servings.

ITALIAN-STYLE VEAL SKILLET

(main dish)

4   veal loin chops (about 1½ pounds total weight), trimmed of all fat
¼   teaspoon salt
⅛   teaspoon black pepper, preferably freshly ground
2   tablespoons olive oil
2   garlic cloves, minced
1   sweet Italian pepper or sweet green pepper, cut into 1-inch squares
1   sweet red pepper, cut into 1-inch squares (if unavailable, substitute a
      sweet green pepper or a sweet Italian pepper)
1   large onion, coarsely chopped
1   cup sliced fresh mushrooms (optional)
1   14- to 16-ounce can tomatoes, including juice
½   cup chopped fresh parsley leaves
¼   cup dry white wine or dry sherry
1   teaspoon dried basil leaves
½   teaspoon dried oregano leaves
1   bay leaf
⅛   teaspoon salt
1   tablespoon cornstarch
¼   cup cold water

Sprinkle the chops with the ¼ teaspoon salt and black pepper.

In a large heavy skillet over medium heat, brown the chops in the oil. Remove the chops from the skillet and set them aside. In the fat remaining in the skillet, cook the garlic, green and red peppers, onion, and mushrooms (if used) just until the onion is tender. Return the chops to the skillet. Add all the remaining ingredients, *except* the cornstarch and cold water, breaking up the tomatoes with a large spoon. Cover and simmer for about 25 minutes, or until the chops are tender. Combine the cornstarch and the cold water and mix well. Move the chops to one side of the skillet. Add the water and cornstarch mixture to the skillet and stir to blend well. Continue stirring until the sauce thickens slightly.

Makes 4 servings.

Potatoes

Although we know the potato as a vegetable with white flesh, it was not always like this. The Peruvian and Bolivian Indians—who were growing potatoes long before Columbus discovered America—cultivated a wide variety of colors ranging from pink and scarlet to royal purple and even blue!

The potato was carried from the New World to Europe by Spanish explorers in the sixteenth century. In the Old World, the plant was initially grown only for its showy flowers and was slow to be accepted by cooks.

The Irish were the first Europeans to appreciate the potato's importance as food. Because the plant adapted easily to growing conditions in the Emerald Isle and could be cultivated with little labor, its popularity spread quickly. By the mid-seventeenth century, it was the country's preeminent food source. This dependence led to disaster when a fungus destroyed virtually the entire Irish potato crop of 1845 and 1846. About two and a half million people died as a result. And the "Potato Famine," as it was known, was a key factor in spurring Irish emigration to America.

Among European countries, France was particularly reluctant to accept the potato—probably because it was thought to be poisonous. (This notion was not unreasonable considering that the potato is related to deadly nightshade.) However, a French military pharmacist named Auguste Parmentier became the potato's advocate after he learned to enjoy the vegetable while imprisoned in Germany during the Seven Years War. Parmentier persuaded Louis XVI to let him grow potatoes in a neglected sandy field. The plants flourished and Parmentier posted a guard over them during the day. Thinking the crop must be valuable, some of the local populace sneaked into the field at night when the guard was off duty and stole the potatoes—just as Parmentier had intended. And he was even more pleased when potatoes began appearing in local gardens. (Today, the word "Parmentier" on a French recipe is an indication that it includes potatoes.)

Initially, the potato was not a hot seller in North America, because it was thought to be "unwholesome." Yet some authorities suggested that it might prove good food for pigs. (Incidentally, it was not brought here directly from South America, but by way of Ireland.)

The first potatoes grown in what became the United States were a small yellow variety called "kidney potatoes." It was not until the late eighteenth century that larger white potatoes were introduced.

Now there are hundreds of white potatoes on the market. But, regardless of variety, they are generally divided into two basic types.

New Potatoes are harvested before they are fully mature. Characteristically, they have thin, sometimes peeling, skin, which is often red or light brown. However, they vary in size, shape, and skin color, depending on the variety. Because new potatoes have a "waxy" texture and hold their shape well, they are good for boiling and in potato salads.

Mature potatoes include both rounded and long varieties. There are four main types, which can be classified according to their suitability for various methods of cooking. The *round white* or "Irish" potato has a thin skin, unobtrusive eyes, and a firm waxy texture, which makes it ideal for boiling. The *round red* is almost identical to the round white, except for its reddish skin. *Baking potatoes* have a fairly thick skin and many eyes. High starch and low moisture content result in their characteristically dry, fluffy texture when baked. They are also excellent mashed and as French fries. The best known variety is the russet, which is also popularly called an Idaho potato. The *long white* is similar in appearance to the russet but is more of an "all-purpose" potato. It can be baked, although it is not quite as fluffy as the russet. It can also be boiled, but is not as satisfactory for this purpose as the round red or white. If you have a choice, buy the potato type that works best for the dish you have in mind.

Availability: Because mature potatoes can be commercially stored for up to 12 months, they are available year round. The peak season for new potatoes is late winter to early spring.

Nutritional Value: Potatoes are a good source of vitamin C and potassium; they also have some calcium, thiamine, and niacin. Contrary to popular belief, they are not really high in calories. A 3½-ounce baking potato has less than 100 calories.

Choosing the Best: Select smooth, firm potatoes. Those with an irregular shape will be harder to peel than more regular ones. Potatoes should be free from cuts, bruises, discolored areas, and decay. Avoid any that have sprouted or shriveled or have green discolorations under the skin. These green areas, called "sunburn," are caused by exposure to light, and they contain a bitter-tasting toxin called solanine. However, the toxin does not spread throughout the whole potato. It affects only those areas of the skin that have turned green, and these can be cut away. Since dirt is sometimes left on potatoes to help retard sunburn, a slightly dirty potato may be a better bet than a clean one with green patches.

Storage: New potatoes should be stored in the refrigerator. Do not leave potatoes in direct sunlight as this causes sunburn. Mature potatoes can be held in a cool, dry place with good ventilation for 2 to 3 months. New potatoes should be used within 2 weeks.

Basic Preparation and Cooking: Potatoes should be scrubbed well before cooking. Cut away any eyes or patches with green discoloration.

For most purposes (even salads and casseroles) there's no need to peel the potatoes unless you prefer to. Much of the nutritional value is found in or just under the skin. Peeled potatoes begin to discolor quickly. If they cannot be used right after they are peeled, cover them with cold water.

Small new potatoes and round red or round white potatoes (up to 2 inches in diameter) can be boiled or steamed whole. Larger ones should be cut into quarters, slices, or cubes. Cooking time depends on size, variety, and the age of the potato. Whole potatoes need to be boiled for 20 to 30 minutes. Slices generally take 8 to 15 minutes. Cubes take 9 to 15 minutes. Steaming will take a few minutes longer. To test for doneness, pierce with a fork. Pieces or whole potatoes should be tender all the way through.

To bake potatoes, scrub them well and dry with a paper towel. Pierce the skin several times with a fork to allow steam to escape. Place the potatoes directly on the oven rack or on a baking sheet. Bake in a preheated 400- to 450-degree oven for 40 to 50 minutes, or until the potatoes feel soft when pressed and are tender all the way through when pierced with a fork. The exact baking time will depend on the potato's size. Potatoes can also be baked in a slower oven if this is more convenient. At 350 degrees, they will take about 1½ hours.

Potatoes bake particularly well in the microwave oven. Pierce them with a fork to allow steam to escape. Microwave 1 potato at full power for about 4 to 6 minutes. To microwave several potatoes, arrange them on a dish with one end pointing toward the center and the other end pointing toward the rim, like the petals of a daisy. For 2 potatoes, microwave 6 to 8 minutes; for 3 potatoes, 8 to 12 minutes; for 4 potatoes, 12 to 16 minutes. Rotate the plate a quarter turn and turn the potatoes over once during cooking. The potatoes may still feel slightly firm at the end of the designated time. Let them stand on the counter for about 5 minutes to soften.

Simple Serving Suggestions: Boiled or baked potatoes can be seasoned with chives, minced parsley leaves, or paprika as well as butter, salt, and pepper.

For mashed potatoes, cut 2½ pounds of russet (Idaho) or long white potatoes (about 4 large) into 1-inch cubes. Put the cubes into a medium-sized saucepan and cover with water. Bring to a boil over high heat. Lower the heat, cover the pot, and simmer the potatoes until they are very tender—about 15 to 17 minutes. Drain in a colander. In the pot in which the potatoes were cooked, mash them with a fork or potato masher until they are completely smooth. Or put them through a food mill. Add about 4 to 5 tablespoons of butter or margarine, ½ to ⅔ cup of hot milk, ¾ teaspoon of salt, and ¼ teaspoon of black pepper. Beat the potatoes with a fork until they are fluffy. Note that the amounts of butter and milk are approximate because potatoes vary in their ability to absorb fats and liquids. Start with the smaller amount called for and use more if necessary. In order to keep the potatoes fluffy, make sure the milk is hot. For a delicious variation, omit 1 or 2 tablespoons of the butter and substitute ½ cup grated Cheddar cheese. Or cook a finely chopped onion with the potatoes.

CRISP POTATO SKIN STRIPS

(hors d'oeuvre)

Unlike most potato skin hors d'oeuvres, these are baked rather than deep-fried, so they contain much less fat. Nevertheless, they are crisp and quite tasty.

8	large, long, and narrow Idaho (russet) potatoes (about 8 to 10 ounces each)
¼	teaspoon salt
	Butter for pan (optional)
3 to 4	tablespoons butter or margarine, melted
	About 2 teaspoons mixed finely chopped dried herbs, such as marjoram, basil, thyme, dillweed, tarragon, and parsley*
3 to 4	tablespoons grated Parmesan cheese

Scrub the potatoes well; then prick them with a fork in several places. Sprinkle the damp skins lightly with the salt; then place the potatoes directly on the rack of a preheated 425-degree oven. Bake them for about 50 to 60 minutes, or until they are tender. Remove the potatoes from the oven and cool them on a wire rack until they can be easily handled.

Use a sharp knife to cut each potato into fourths lengthwise; then scoop out most of the flesh, leaving a shell about ⅛ inch to ¼ inch thick. (Set aside the scooped-out flesh for another use.) Use a knife, kitchen shears, or scissors to cut each piece of potato shell in half lengthwise, so that you have a total of 8 strips from each potato.

Place the strips, skin side down, in one layer on a large buttered (or nonstick spray-coated) baking sheet. Bake in a preheated 375-degree oven for 15 minutes. Remove from the oven and brush the strips with the melted butter. Then sprinkle them with your choice of herbs and the grated cheese. Return the strips to the oven and bake them for 12 to 15 minutes longer, or until they are crisp and lightly browned. Cool the potato strips slightly; then serve them warm or at room temperature. (They may be frozen and reheated, if desired.)

Makes 64 strips.

*A good, premixed herb combination that contains no added salt is Bouquet Garni, made by Spice Islands, a division of Specialty Brands, Inc.

CURRIED POTATO-PEANUT SALAD

This unusual combination produces a salad that is healthful as well as tasty. Unsalted peanuts (available at health food stores and some supermarkets) are preferable, but the dry roasted salted ones will also work. Just adjust the dressing seasonings to account for the extra salt.

    2   pounds small "new" red potatoes, scrubbed
    1   cup diced sweet green pepper
    ¾   cup finely chopped scallions, including green tops
    ½   cup diced celery
    ½   cup peeled, seeded, and diced cucumber
    2   tablespoons finely chopped fresh parsley leaves
    ½   cup roasted peanuts, divided

Dressing

        ¾   cup plain lowfat yogurt
        2   tablespoons mayonnaise
        2   tablespoons apple cider vinegar
        2   tablespoons peanut butter
    1 to 1½   teaspoons curry powder (to taste)
            Salt to taste

Put the potatoes in a saucepan with about 1½ inches of water. Cover the pan and bring the potatoes to a boil over high heat. Lower the heat and steam the potatoes until they are fork-tender, about 25 to 35 minutes. Drain off all the water and let the potatoes cool until they can be easily handled; then cut them into ¾-inch cubes. Put the potatoes in a medium-sized bowl with the green pepper, scallions, celery, cucumber, parsley, and all but 2 tablespoons of the peanuts.

Mix all the dressing ingredients in a small bowl; then toss the dressing with the potato mixture. Cover and refrigerate the salad for several hours, stirring occasionally, until it is chilled and the flavors are blended. Before serving, coarsely chop the remaining peanuts and sprinkle them on top.

Makes 6 to 8 servings.

MASHED POTATO CASSEROLE

(side dish)

6   cups uncooked peeled 1-inch potato cubes
1   medium-sized onion, finely chopped
6   ounces mild Cheddar cheese, grated (1½ cups packed)
3   tablespoons butter or margarine
¾   cup instant nonfat dry milk powder
¼   cup plain lowfat yogurt
½   cup water
¼   teaspoon black pepper, preferably freshly ground
½   teaspoon salt

Combine the potatoes and the onion in a large saucepan. Add water to cover. Cover and bring to a boil over high heat. Lower the heat and simmer for about 15 to 17 minutes, or until the potatoes are very tender. Drain the potatoes well in a colander. Transfer them to a large bowl and mash them with an electric mixer on low speed or a potato masher until very smooth. Add all the remaining ingredients and stir with a large spoon until well combined.

Turn the mixture into a lightly greased or nonstick spray-coated 2-quart casserole and bake, uncovered, in a preheated 350-degree oven for about 25 minutes, or until the casserole is heated through.

Makes about 6 servings.

RICOTTA-POTATO CASSEROLE

(side or light main dish)

About 2 pounds small unpeeled "new" red potatoes, scrubbed
2   cups part-skim ricotta cheese (if unavailable, substitute regular ricotta)
1   large egg
2   tablespoons grated or minced onion
2   tablespoons finely chopped fresh parsley leaves
½   teaspoon dried thyme leaves
½   teaspoon dried basil leaves
¼   teaspoon salt
⅛   teaspoon black pepper, preferably freshly ground
¼   cup white or whole wheat fresh bread crumbs (made in a food processor or blender)
1   ounce hard or semisoft cheese, such as mozzarella, Muenster, Swiss, mild Cheddar, etc., grated (¼ cup packed)

Put the potatoes into a medium-sized saucepan with about 1 inch of water. Bring to a boil over high heat. Cover the pan tightly, lower the heat, and steam the potatoes until they are tender, about 25 to 35 minutes. Drain the potatoes well and cool them until they can be handled. Cut them into ⅜-inch-thick slices.

In a medium-sized bowl, mix together the ricotta, egg, onion, parsley, thyme, basil, salt, and pepper.

Layer some of the potatoes in the bottom of a greased or nonstick spray-coated 6-cup casserole. Cover them with a layer of the ricotta mixture; then sprinkle with some of the bread crumbs. Repeat the layers, using up all the ingredients and ending with a layer of potatoes. Sprinkle the grated cheese on top.

Bake the casserole, covered, in a preheated 350-degree oven for 15 minutes. Uncover and bake an additional 15 to 20 minutes, or until hot and bubbly.

Makes about 6 servings.

EASY BROCCOLI-CHEESE-TOPPED BAKED POTATOES

(light main dish or side dish)

There could hardly be an easier way to dress up baked potatoes, yet this topping tastes great.

Topping
2	10-ounce packages frozen chopped broccoli, thawed
⅓	cup commercial sour cream
⅓	cup plain lowfat yogurt
1	teaspoon instant minced onions
¼	teaspoon salt
⅛	teaspoon black pepper, preferably freshly ground

To Serve
5 to 6	large, hot, freshly baked potatoes
6	ounces sharp Cheddar cheese, shredded or grated (1½ cups packed)

Drain the broccoli well in a colander; then fluff it up with a fork. Put the broccoli and all the remaining topping ingredients in a blender or a food processor fitted with a steel blade. Blend or process the mixture until puréed, but not completely smooth; if necessary, stop the blender and stir to redistribute the contents several times.

Transfer the purée to a medium-sized saucepan over medium heat. Heat the purée, stirring frequently, until it is piping hot BUT NOT boiling; serve immediately.

To serve, split the baked potatoes in half lengthwise, and fluff up the insides slightly with a fork. Spoon the purée over the potatoes, dividing it equally among them. Sprinkle the cheese over the broccoli purée. Heat the potatoes in a 300-degree oven for 2 to 3 minutes, or until the cheese has melted. (Alternatively, place the potatoes on a plate in a microwave oven and cook on full power for about 2 minutes, or until the cheese has melted.)

Makes 5 to 6 servings.

EASY DAIRY TOPPING FOR BAKED POTATOES

(side dish topping)

While this topping is much lower in calories than either butter or plain sour cream, it doesn't skimp on taste appeal.

½   cup commercial sour cream
½   cup plain lowfat yogurt
1   teaspoon instant minced onions
¾   teaspoon celery salt
½   teaspoon dried dillweed
⅛   teaspoon black pepper, preferably freshly ground

In a small bowl, mix all the ingredients together. Cover and refrigerate for at least 1 hour to allow the flavors to blend. Serve on baked potatoes as you would sour cream.

Makes about 1 cup.

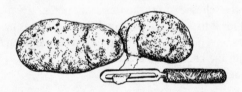

BEEF AND VEGETABLE-TOPPED BAKED POTATOES

(main dish)

This flavorful beef and vegetable combination is designed to be served over piping hot baked potatoes. Each potato should be cut open and the insides fluffed up slightly with a fork before the mixture is spooned on top. Other simple toppings, such as grated Cheddar cheese, sour cream, and chopped chives, or Easy Dairy Topping (page 216) or Broccoli-Cheese Topping (page 215) can be offered along with this one so that diners can mix and match. If you'd like to make more servings of this particular topping, simply double the recipe.

¾   pound lean ground beef
1   medium-sized onion, finely chopped
⅓   cup diced sweet green pepper
1   celery stalk, thinly sliced
1   garlic clove, minced
1   15-ounce can tomato sauce
¼   teaspoon salt
⅛   teaspoon black pepper, preferably freshly ground
1   bay leaf
½   teaspoon dried thyme leaves
½   teaspoon powdered mustard
½   teaspoon dried marjoram leaves
2 to 3   drops Tabasco sauce
½   teaspoon sugar

To Serve
3 to 4   large, hot, freshly baked potatoes

In a medium-sized saucepan over medium-high heat, combine the ground beef, onion, green pepper, celery, and garlic, breaking up the meat with a spoon, until the meat is brown and the onion is tender. Drain off and discard any excess fat. Add all the remaining topping ingredients and stir to blend well. Bring to a boil. Lower the heat, cover, and simmer for about 25 minutes, stirring occasionally, or until the flavors are well blended and the vegetables are tender. Serve over baked potatoes.

Makes 3 to 4 servings.

SHEPHERDS PIE

(main dish)

Hearty and satisfying, this mashed potato-topped stew makes a fine one-dish meal for a winter day. A flameproof and ovenproof casserole may be used instead of the skillet and casserole called for.

Stew

2	tablespoons vegetable oil
1¼	pounds lean stew beef, cut into 1-inch cubes
1	large onion, finely chopped
1	garlic clove, minced
2	medium-sized carrots, thinly sliced
2	celery stalks, thinly sliced
3	8-ounce cans tomato sauce
2	cups 2-inch fresh green bean pieces, stem ends removed
2	bay leaves
½	teaspoon dried marjoram leaves
½	teaspoon dried thyme leaves
¼	teaspoon black pepper, preferably freshly ground
½	teaspoon salt
¼	teaspoon powdered mustard
2 to 3	drops Tabasco sauce

Topping

4	cups 1-inch uncooked peeled potato cubes
⅓	cup lowfat or regular milk
3	tablespoons butter or margarine, slightly softened
¼	teaspoon salt
⅛	teaspoon black pepper, preferably freshly ground

In a large heavy skillet over medium-high heat, heat the oil. Then brown the meat on all sides, stirring frequently. Add the onion and garlic. Continue cooking, stirring, until the onion is tender. Add the carrots and celery and cook for 2 to 3 minutes longer. Add the remaining stew ingredients. Stir to blend well. Bring to a simmer. Transfer the stew to a 2½-quart casserole. Bake, covered, in a 350-degree oven for about 50 minutes, or until the vegetables are almost tender.

Meanwhile, for the topping, combine the potatoes, and enough water to cover, in a large saucepan. Cover the saucepan and bring to a boil over high heat. Lower the heat and simmer for about 15 to 17 minutes, or until the potatoes are very tender. Drain the potatoes well in a colander. Return them to the cooking pot, along with the remaining topping ingredients. Mash with a potato masher. Remove the casserole from the oven and uncover. Drop the potato topping by large spoonfuls onto the bubbling mixture. Carefully spread out the potatoes with the back of the spoon. Bake, uncovered, for 15 minutes, or until the potatoes are heated through.

Makes about 6 servings.

LIGHT WHEAT-POTATO-YEAST BREAD

Potatoes add flavor and moistness to this good all-purpose loaf bread. The loaves are large and pale tan in color. The recipe can also be made fairly quickly, since the new fast-rising yeast is used.

1	large boiling potato, peeled and cut into ½-inch cubes
4½ to 5	cups enriched all-purpose white or unbleached flour
2	packets fast-rising dry yeast
⅓	cup sugar
¼	cup instant nonfat dry milk powder
2	teaspoons salt
¼	cup butter or margarine, cut into 3 or 4 pieces
1	large egg
	Reserved potato cooking water
1½	cups whole wheat flour

Put the potato cubes in a small saucepan and just cover them with water. Bring them to a boil over medium heat and simmer, covered, for 8 to 10 minutes, or until they are just tender.

Meanwhile, stir together 1 cup of the white flour, the yeast, sugar, milk powder, and salt in a bowl. Set aside.

Drain the potato cubes well, reserving the cooking liquid in a large measuring cup. Transfer the potatoes to a large mixer bowl. With the mixer on low speed, beat for about 1 minute, or until the potatoes are well mashed. Scrape down the sides of the bowl. Increase the speed to medium and beat for about 1 minute longer, or until the potatoes are completely smooth. Add the butter and continue beating for 1 minute; then add the egg and beat for 30 seconds longer. Add enough water to the reserved potato liquid to make 1½ cups. If necessary, return the water-potato liquid mixture to a small saucepan and heat to 125 to 130 degrees (hot to the touch but not scalding).

Add the flour-yeast mixture and then the water-potato liquid mixture to the mixer bowl. Beat with the mixer on low speed until the ingredients are blended. Increase the speed to medium and beat for 5 minutes. Vigorously stir in all of the whole wheat flour, then 3 cups of the white flour with a large spoon. Working in the bowl or on a clean, lightly floured surface, vigorously knead in about ½ to 1 cup of additional white flour, or enough more to yield a smooth and malleable, yet still slightly moist, dough. Form the dough into a ball and transfer it to a large greased bowl. Cover the bowl with plastic wrap and set aside in a very warm spot for 25 to 30 minutes, or until the dough is doubled in bulk.

Punch down the dough and divide it in half. With well-greased hands, shape the dough into loaves and place in 2 well-greased 9- by 5- by 3-inch loaf pans. Cover the pans with plastic wrap and set them aside in a very warm spot for about 15 minutes, or until doubled in bulk. Carefully cut a ¼-inch-deep slash lengthwise down the center of each loaf with a sharp knife.

Bake in a preheated 400-degree oven for 30 to 35 minutes, or until the loaves are nicely browned and sound hollow when tapped on the bottom. Remove the loaves from the pans and transfer them to wire racks. The bread is good served still warm from the oven, although it slices better if it is allowed to cool thoroughly. It may also be used for toast.

Makes 2 loaves.

Rutabagas

Though they are often called "yellow turnips" or "Swedish turnips," rutabagas actually differ in many ways from their botanical cousins. While the turnip is an ancient vegetable, the rutabaga did not appear until the seventeenth century. Moreover, rutabagas are generally much larger than turnips and more bulbous, with a slightly elongated, ridged "neck" or "crown."

The rutabaga also has a stronger, somewhat sweeter flavor than the turnip, and stores much better due to a lower water content and denser flesh. And, the green tops and the skins of rutabagas are never eaten as they sometimes are with young turnips.

Rutabaga plants are slow growers that prefer a cool climate, and will not thrive where the summers are exceedingly hot. For this reason, most of the American supply comes from Canada. Rutabagas also grow well in Scandinavia, and are particularly popular in Sweden, which explains why most English-speaking Europeans call them "swedes." (However, the Scots sometimes call rutabagas "neeps," an English nickname for white turnips, which adds to the confusion between the two vegetables.) The common American name derives from the Swedish *rotbagga* or *rotobagge*, which means "ram's root."

Rutabagas probably originated in Bohemia, and may have been the result of cross-breeding cabbages and turnips. Genetic evidence for this comes from chromosome counts: the cabbage has 18 chromosomes, and the turnip, 20; whereas the rutabaga has 38—equal to the other two combined.

Commercially available rutabagas almost always have brownish-yellow to pale orange skin, tinged with violet near the neck. However, this root vegetable also comes in other colors, such as white and purple. The skin is often coated with paraffin to help preserve moisture and prevent shriveling, thus increasing shelf life. Rutabaga flesh is usually pastel orange or dark yellow.

Availability: In many areas, rutabagas are available year round; however, they are most frequently seen from July through April, with a peak from October through March.

Choosing the Best: Although size does not generally affect quality, excessively large rutabagas may not be as flavorful as smaller ones. Rutabagas should be heavy for their size and very firm. The skin should be relatively smooth, with no bruises or signs of decay.

Nutritional Value: Rutabagas are a good source of both vitamins A and C (and contain much more of these nutrients than turnips). Rutabagas also have moderate amounts of several minerals, including potassium, calcium, and magnesium.

Storage: Rutabagas need cool temperatures and high humidity. Refrigerate them in the vegetable crisper or a loosely sealed plastic bag. Depending on whether or not they are waxed, they will keep from about a week up to 1 month or longer.

Preparation and Basic Cooking: Rutabagas must be peeled before eating. Use a sharp knife to peel and cut them into chunks, slices, strips, cubes, or dice. To cook, boil or steam the pieces until tender. Boiled ½-inch cubes will take about 15 to 20 minutes; steamed ones about 10 minutes longer. A teaspoon of sugar added to the cooking water may improve the flavor of boiled rutabaga.

Another method is to sauté small pieces of rutabaga in a little butter or margarine; then add a small amount of water or broth, cover, and complete the cooking by braising for about 15 to 20 minutes, or until the rutabaga is tender.

Larger rutabaga chunks may be baked or roasted—a good way to cook them is with meat. In a 350-degree oven, 1-inch chunks will take about 40 to 50 minutes.

Simple Serving Suggestions: Thinly sliced raw rutabaga makes a nice, change-of-pace crudité to serve with dips. Add julienne strips or shredded raw rutabaga to salads. Boiled or steamed rutabaga pieces taste good with butter. They are particularly tasty when mashed and seasoned like sweet potatoes. Rutabagas are also very good in stews and New England-style boiled dinners, as they absorb some of the flavor of the meat with which they are cooked.

HEARTY BEEF AND VEGETABLE SOUP

 9 cups water
 3 pounds meaty beef soup bones
 1 pound lean stew beef, cut into 1-inch cubes
 ½ cup dry navy beans, sorted and washed
 ¼ cup pearl barley
 2 medium-sized carrots, coarsely sliced
 2 celery stalks, including leaves, coarsely sliced
 2 large onions, coarsely chopped
 1 garlic clove, minced
 1 15-ounce can tomato sauce
 ½ cup chopped fresh parsley leaves
 3 bay leaves
1½ teaspoons celery salt
 1 teaspoon dried thyme leaves
 1 teaspoon dried basil leaves
 1 teaspoon powdered mustard
 ¾ teaspoon chili powder
 1 teaspoon salt
 ¾ teaspoon black pepper, preferably freshly ground
 2 cups diced peeled rutabagas
 3 cups 2-inch fresh green bean pieces, stem ends removed
 2 cups thinly sliced cabbage
 2 cups fresh or frozen loose-pack corn kernels

In a large heavy pot over high heat, combine the water, soup bones, meat, beans, barley, carrot, celery, onion, garlic, tomato sauce, parsley, and seasonings. Bring to a boil. Cover, lower the heat, and simmer for 1½ hours, stirring occasionally. Remove the soup bones. Turn off the heat. With a large spoon, carefully skim off and discard the fat from the top of the soup.

Add the rutabagas, green beans, cabbage, and corn. Over medium-high heat, return the soup to a boil, stirring occasionally. Lower the heat and simmer, stirring occasionally, an additional 40 minutes, or until the rutabagas and green beans are tender. Meanwhile, cut the meat from the soup bones, trim off the fat and discard it. Return the meat to the soup pot and heat through.

Makes 9 to 12 servings.

SPICY RUTABAGA AND ONION

(side dish)

1   tablespoon butter or margarine
1   small onion, finely chopped
1   cup beef bouillon (reconstituted from cubes or granules)
⅛   teaspoon ground ginger
⅛   teaspoon ground allspice
⅛   teaspoon black pepper, preferably freshly ground
    Pinch of ground thyme
3   cups ½-inch peeled rutabaga cubes

In a small saucepan, melt the butter over medium-high heat. Add the onion and cook, stirring frequently, until it is tender. Add the bouillon, ginger, allspice, pepper, and thyme. Stir to combine well. Add the rutabaga. Cover, lower the heat, and simmer for about 15 minutes, or until the rutabaga is almost tender. Uncover the saucepan, raise the heat, and rapidly cook down the liquid, stirring often, until about ⅓ cup remains as a sauce.

Makes about 4 servings.

MASHED RUTABAGA AND POTATO

(side dish)

3   cups ½-inch peeled rutabaga cubes
3   cups ½-inch peeled potato cubes
2   tablespoons butter or margarine
¼   cup lowfat or regular milk
¼   teaspoon salt
⅛   teaspoon black pepper, preferably freshly ground

Put the rutabaga cubes in a medium-sized saucepan and cover with water. Cover and bring to a boil over high heat. Lower the heat and simmer the rutabaga for about 25 minutes. Add the potatoes to the pan and continue cooking for 10 minutes longer, or until the potatoes and the rutabaga are tender. Drain well in a colander. Transfer the rutabaga and the potatoes to a medium-sized bowl and mash them with a potato masher or an electric mixer on low speed. Add all the remaining ingredients. Blend well.

Makes about 4 servings.

SPICED RUTABAGA AND APPLE CASSEROLE

(side dish)

Although rutabagas are related to turnips, they are actually much sweeter. In this dish, they look and taste a bit like sweet potatoes or winter squash.

5   cups peeled and diced rutabagas (about 2¼ pounds)
1   cup water
1   tablespoon honey
3   cups peeled and finely diced apples (3 large ones)
1   tablespoon lemon juice, preferably fresh
3   tablespoons packed dark brown sugar, divided
¼   teaspoon ground cinnamon
⅛   teaspoon ground nutmeg
2   tablespoons butter or margarine, divided
¼   cup bread crumbs or dry unsweetened breakfast cereal, such as corn flakes

In a 2-quart saucepan, combine the rutabagas, water, and honey. Cover the pan tightly and bring to a boil over medium-high heat. Lower the heat and simmer the rutabagas about 25 minutes, or until they are very tender.

Meanwhile, mix the apples with the lemon juice, 1 tablespoon of the brown sugar, the cinnamon, and nutmeg. Set aside.

When the rutabagas are tender, drain them well. Transfer the rutabagas to a bowl and mash them coarsely. Stir in 1 tablespoon of the butter until it is melted. Then add the remaining 2 tablespoons of brown sugar.

Mix the seasoned apples with the rutabagas. Transfer the mixture into a lightly greased or nonstick spray-coated 6-cup casserole. Sprinkle the bread crumbs on top; then dot with the remaining 1 tablespoon of butter, cut into small pieces. Cover the dish with a lid or foil. Bake in a preheated 350-degree oven for 35 minutes. Then remove the cover and continue baking for about 10 minutes longer, or until the apples are very tender.

Makes 6 to 8 servings.

SAVORY WINTER VEGETABLE BAKE

(side dish)

Sauce
2½   tablespoons butter or margarine
1   tablespoon finely chopped fresh chives or scallion
1   tablespoon enriched all-purpose or unbleached white flour
¾   cup water
2   tablespoons Dijon or Dijon-style mustard
1   teaspoon sugar
¼   teaspoon dried tarragon leaves
¼   teaspoon celery salt
⅛   teaspoon black pepper, preferably freshly ground

Vegetables
2   medium-sized rutabagas, peeled and cut crosswise into ⅛-inch-thick slices
1   medium-sized red onion, cut crosswise into ⅛-inch-thick slices
3   medium-sized carrots, cut diagonally into ¼-inch-thick slices
4 to 5   cabbage wedges, each ¾ inch thick at the thickest part

Combine the butter and chives in a small saucepan over medium-high heat. Cook for 2 to 3 minutes, stirring occasionally. Stir in the flour until well blended and smooth; cook, stirring, for 2 minutes longer. Stirring vigorously, gradually add the water and then all the remaining sauce ingredients. Continue to cook for about 2 minutes, or until the sauce is slightly thickened and completely smooth. Set aside.

Put half the rutabaga slices in the bottom of a 1½- to 2-quart ovenproof casserole or Dutch oven. Top the rutabaga with about a third of the onion slices. Drizzle the vegetable layer with about a fourth of the sauce. Add the carrots, then another third of the onions to the casserole. Drizzle with another fourth of the sauce. Add the remaining rutabaga and then the onions. Top with another fourth of the sauce.

Bake, tightly covered, in a preheated 375-degree oven for 25 minutes. Lay the cabbage wedges over the other vegetables and top with the remaining sauce. Return the casserole to the oven and bake for about 25 minutes longer, or until the vegetables are just tender.

Makes 4 to 5 servings.

Spinach

Of all the various vegetables classified as "greens," spinach is perhaps the most widely known and enjoyed. The leafy, low-growing plant was probably first cultivated about 2,000 years ago in the area now called Iran. By the seventh century A.D., spinach could be found as far east as China, where it was called the "herb of Persia." Several centuries later, it had also spread westward to Spain, most likely introduced there by the Moors.

By the Middle Ages, this tender-leafed member of the goosefoot family was used for food throughout Europe. Spinach was a common crop in monastery gardens on the continent and was enjoyed in the court of the English king Richard II as well.

It is uncertain when Popeye's favorite vegetable was introduced into the United States, but three different varieties were flourishing by about 1800. At that time, only smooth-leafed types existed; the crinkly, or "Savoy-leafed," spinach did not appear until 1828. Due to its springiness and appealing ruffled look, this more modern type is favored by retailers selling fresh spinach, but the smooth-leaf kind is more widely grown for use by canners and frozen food processors. The two taste essentially the same.

Availability: A fairly constant supply of fresh spinach is available year round. Frozen spinach, which makes a very acceptable and convenient substitute for fresh in some recipes, is also widely sold. Many supermarkets stock canned spinach as well, although it lacks the delicate taste and bright color of fresh or frozen.

Choosing the Best: When buying fresh spinach, look for leaves that are bright green, lustrous, and unblemished. Also be sure the leaves are fairly free of grit. Additionally, the "Savoy," or crinkly-leafed, type should be crisp and somewhat "bouncy" to the touch. If purchasing prepackaged spinach, be wary of excessive condensation inside the bag, as this strongly suggests the contents are deteriorating.

Nutritional Value: Particularly when eaten raw, spinach is an excellent source of vitamin A and potassium, and a good source of vitamin C. Although the vegetable contains a fair amount of calcium, it is of no nutritional benefit because it combines with other substances in the plant (called oxalates) to form a compound the human body can't absorb.

Storage: It is important to use spinach as soon after purchase as possible. If absolutely necessary, it can be held in the refrigerator loosely packed in a plastic bag for a short period. However, it is best not to store spinach longer than about 24 hours. Lengthy storage not only adversely affects the taste of spinach, but leads to the formation of nitrites—chemicals that are potentially carcinogenic.

Preparation and Basic Cooking: Thoroughly rinse spinach in a colander under cool running water to remove any superficial grit. Using a sharp knife, trim off the stems and tough midribs. Then turn out the leaves into a sink of barely warm water, and gently swish them around to loosen caked-on soil or sand trapped in the crinkles. Drain off the water and wash out the sink; then refill with warm water and wash the leaves again. Keep repeating the process until the water looks clear and no silt remains in the bottom when the sink is drained. Return the spinach to a colander; rinse once more and drain well. Dry the leaves by whirling in a salad spinner or blotting on paper towels. They are now ready to be used raw or cooked.

To cook fresh spinach that has been blotted dry and stored, combine it in a very large pot with a teaspoon or two of water. For spinach that still has water clinging to its leaves, no additional cooking liquid is necessary. Cook, stirring, for 3 to 5 minutes, or until the leaves are reduced in volume and just tender. Spinach may also be steamed in a large steamer basket or colander over boiling water for 7 to 10 minutes.

Simple Serving Suggestions: Dress plain-cooked spinach with butter, salt, and pepper. A dash of lemon juice may be added, if desired. For a different taste, stir 2 or 3 tablespoons of cream into boiled or steamed spinach just before serving. Or sprinkle spinach with grated Parmesan cheese. Raw spinach is delicious in tossed salads.

SAUTÉED SPINACH WITH HERBS

(side dish)

This dish is simplicity itself, yet quite tasty. Cooking time is also very short—less than 5 minutes.

2	tablespoons butter or margarine
1	small garlic clove, minced
1	tablespoon chopped fresh chives (or 1½ teaspoons dried chopped chives)
1½	pounds fresh spinach leaves, washed (see directions under Preparation), stemmed and torn into bite-sized pieces
⅛	teaspoon dried basil leaves
	Pinch of dried thyme leaves
	Pinch of crumbled dried rosemary leaves
	Generous ⅛ teaspoon salt
⅛	teaspoon black pepper, preferably freshly ground
1	tablespoon grated Parmesan cheese

Melt the butter in a very large skillet or sauté pan over medium-high heat. Add the garlic and chives and cook, stirring, for 1 minute. Gradually stir in the spinach leaves; at first they will overfill the pan but will rapidly decrease in volume. (If the leaves are dry, add a teaspoon of water to the pan along with them.) Stir in all the remaining ingredients, *except* the Parmesan cheese, until well mixed. Sauté the spinach, stirring, for 3 to 5 minutes, or until the leaves are just tender. Sprinkle the spinach with the Parmesan cheese and serve.

Makes 4 to 5 servings.

NO-FAIL SPINACH-CHEESE SOUFFLÉ

(side dish or light main dish)

This soufflé is practically foolproof because the Swiss or Gruyère cheese helps to give it body so that it holds its shape well. As all cheeses do not work in this same way, it is best not to use substitutes in this recipe.

5   ounces Swiss or Gruyère cheese, finely grated (1¼ cups packed), divided
2   tablespoons butter or margarine
¼   cup finely chopped onion
3   tablespoons enriched all-purpose or unbleached white flour
1   cup hot tap water
⅓   cup instant nonfat dry milk powder
3   large eggs, separated
⅛   teaspoon salt
    Scant ⅛ teaspoon black pepper, preferably freshly ground
    Pinch of ground nutmeg
1   10-ounce package frozen chopped spinach, thawed and squeezed dry of
    all excess liquid
2   large egg whites
¼   teaspoon cream of tartar

Lightly grease or coat with nonstick vegetable spray a 1½-quart soufflé dish or similar straight-sided casserole. Sprinkle the bottom and sides with about 2 tablespoons of the cheese.

In a medium-sized saucepan over medium-high heat, melt the butter; then cook the onion, stirring, until it is tender but not browned. Add the flour and cook, stirring, for 1 minute. Use a wire whisk to gradually stir in the hot water and then the milk powder. Continue to heat and stir vigorously until the sauce has thickened considerably and boiled for 1 minute. Remove the sauce from the heat and beat in the egg yolks, one by one, and then the salt, pepper, and nutmeg. Use a spoon to stir in the spinach and all but 2 tablespoons of the remaining cheese.

In a large bowl, use an electric mixer or a wire whisk to beat all 5 egg whites with the cream of tartar just until they are stiff but not dry. Stir about a fourth of the whites into the spinach mixture to lighten it. Then gently but thoroughly fold the lightened spinach mixture into the remaining whites. Transfer the mixture to the prepared dish. Sprinkle the reserved 2 tablespoons of cheese on top.

Bake the soufflé in a preheated 375-degree oven for about 30 to 35 minutes, or until it is puffed and seems set in the center; that is, it does not jiggle excessively. (The baked soufflé will "hold" in a hot, turned-off oven for about 5 minutes, if necessary.) Serve the soufflé as soon as possible, as it will begin to deflate a few minutes after it is removed from the oven. Use a large serving spoon to dish out portions, including some "crust" with each portion.

Makes 4 to 6 side-dish servings or 2 to 3 main-dish servings.

GREEK-STYLE SPINACH "PIE"

(main dish)

Although this version of Spinach Pie tastes quite traditional, it features not only spinach, but also kale, which adds flavor as well as abundant nutrients. In addition, the recipe calls for much less fat than most similar dishes.

This pie is made with filo, a very thin, low-calorie dough that looks and feels remarkably like white tissue paper. The delicate sheets of filo (also spelled "fillo" or "phyllo") dry out very quickly; therefore, they should be kept in a stack under a damp cloth once they are unwrapped. One-pound packages of filo sheets (sometimes called "strudel leaves") can be purchased at some supermarkets as well as most Greek or Middle Eastern ethnic groceries and gourmet specialty stores. If the dough has been frozen, it should be thawed completely in the refrigerator and then brought to room temperature before being unwrapped and used. Leftover dough should be tightly re-wrapped and refrigerated or frozen.

2	tablespoons olive or vegetable oil
1	medium-sized onion, finely chopped
4	scallions, including green tops, thinly sliced
2	10-ounce packages frozen chopped spinach, thawed and squeezed dry
1	10-ounce package frozen cut leaf kale, thawed and squeezed dry
1	15-ounce carton part-skim ricotta cheese (if unavailable, substitute regular ricotta)
½	pound feta cheese, crumbled
2	large eggs plus 2 large egg whites
2	tablespoons finely chopped fresh parsley leaves
1 to 2	teaspoons dried dillweed (to taste)
	Generous ⅛ teaspoon black pepper, preferably freshly ground
⅛	teaspoon salt (or to taste)
3½	tablespoons butter, preferably unsalted (if unavailable, substitute margarine)
2	teaspoons vegetable oil
2	teaspoons water
	About ½ pound packaged filo (or "strudel leaves")

In a large skillet over medium-high heat, heat the olive oil and cook the onion and scallions, stirring, until they are tender but not browned. Add the spinach and kale and cook, stirring, until any excess moisture has evaporated. Remove the skillet from the heat to cool the spinach-kale mixture slightly.

In a large bowl, stir together the ricotta, feta cheese, eggs, egg whites, parsley, dillweed, pepper, and salt. Stir in the spinach-kale mixture until well combined. Adjust the seasonings to taste. Set aside.

In a small saucepan over medium-low heat, heat the butter with the 2 teaspoons of oil; then stir in the water. Turn the heat to the lowest setting, just to keep the mixture warm. Brush some of the butter mixture in the bottom of a 9- by 13-inch (or equivalent) baking pan. Using kitchen shears, trim several sheets of filo so they will fit in the pan. (With some brands of filo, the large sheets can simply be cut in half to fit perfectly into a 9- by 13-inch pan.) Discard any filo trimmings or reserve them for another use.

Carefully place 1 trimmed sheet of filo in the bottom of the buttered pan. Sprinkle or very lightly brush (with a pastry brush) a small amount of the butter mixture on top; then cover with another sheet and more butter. Continue until 8 sheets of filo have been layered. (Don't worry if a sheet tears; just put the piece together for a layer.)

Spoon all the spinach-kale filling on top, spreading it evenly. Cover the filling with another trimmed sheet of filo; then sprinkle or brush with the butter mixture as before. Continue layering filo until all the butter mixture is used up, making sure that the last sheet on top is well buttered. There should be at least 8 layers of filo in the top crust. (The pie may be assembled to this point and refrigerated for several hours or overnight, if desired.)

Bake the pie in a preheated 350-degree oven for about 45 to 55 minutes, or until the top is lightly browned and crisp. To serve the pie, cool it slightly; then cut it into large squares with a sharp knife.

Makes about 6 servings.

SPINACH-PASTA MAIN-DISH CASSEROLE

(main dish)

2½ cups fusilli (curly pasta) or spiral or shell-shaped pasta
1 10-ounce package frozen chopped spinach, thawed and drained
1 15-ounce carton part-skim ricotta cheese (if unavailable, substitute regular ricotta)
½ cup grated Parmesan cheese, divided
½ teaspoon dried marjoram leaves
¼ teaspoon ground nutmeg
¼ teaspoon salt
⅛ teaspoon black pepper, preferably freshly ground

Sauce
1 tablespoon olive or vegetable oil
1 small onion, chopped
⅓ cup chopped fresh parsley leaves
1 garlic clove, minced
1 16-ounce can tomatoes, including juice
1 6-ounce can tomato paste

1½ teaspoons dried oregano leaves
½ teaspoon dried marjoram leaves
¼ teaspoon celery seeds
¼ teaspoon salt
⅛ teaspoon black pepper, preferably freshly ground
    Scant ⅛ teaspoon crushed hot red pepper (optional)

Cook the pasta according to package directions (in unsalted water) until almost tender. Add the spinach; if necessary separate the pieces with the tines of a fork. Continue cooking until the water just returns to a boil. Immediately transfer the pasta-spinach mixture to a colander and set it aside to drain.

Stir together the ricotta, ¼ cup of the Parmesan cheese, marjoram, nutmeg, salt, and pepper in a small bowl.

Prepare the sauce as follows: Combine the oil, onion, parsley, and garlic in a medium-sized saucepan. Cook, stirring occasionally, over medium-high heat for 4 to 5 minutes, or until the vegetables are limp. Stir in the tomatoes, breaking them up with a spoon. Stir in the tomato paste and all the remaining sauce ingredients. Lower the heat and simmer the mixture, covered, for 5 minutes.

To assemble the casserole, stir together the spinach-pasta mixture and the ricotta mixture and put it into a 2½- to 3-quart ovenproof casserole. Spoon the sauce evenly over the top. Sprinkle the surface with the remaining ¼ cup Parmesan cheese.

Bake, uncovered, in a preheated 375-degree oven for 30 to 35 minutes, or until the casserole is heated through and bubbly.

Makes 5 to 6 servings.

SPINACH LASAGNE

(main dish)

Sauce
1 tablespoon butter or margarine
2 garlic cloves, minced
1 large onion, finely chopped
1 16-ounce can tomatoes, including juice, puréed in a blender or food processor
1 10¾-ounce can tomato purée
1 6-ounce can tomato paste
1 broccoli stem, grated or very finely chopped (Reserve the flowerets for another use.)
1 medium-sized carrot, grated or very finely chopped
1 celery stalk, finely chopped
1 teaspoon dried basil leaves

1¼   teaspoons dried oregano leaves
 1   bay leaf
 ¼   teaspoon black pepper, preferably freshly ground
     Scant ½ teaspoon salt

Noodles and Filling
 15   lasagne noodles
  1   15-ounce carton part-skim ricotta cheese (if unavailable, substitute
      regular ricotta)
  1   10-ounce package frozen chopped spinach, thawed and very well
      drained
  8   ounces mozzarella cheese, grated (about 2 cups packed)
  ⅓   cup grated Parmesan cheese

In a medium-sized saucepan, prepare the sauce: Combine the butter, garlic, and onion. Cook over medium-high heat until the onion is soft. Add all the remaining sauce ingredients. Stir to mix well. Lower the heat and simmer the sauce, covered, for about 20 minutes; then remove the bay leaf. Meanwhile, cook the lasagne noodles according to the package directions.

To assemble the lasagne, line the bottom of a nonstick spray-coated or lightly greased 9½- by 13-inch baking pan with 5 noodles, overlapping them slightly at the edges. Spread with half of the ricotta, all of the spinach, a third of the mozzarella, and a third of the sauce. Follow with another layer of noodles, the remaining ricotta, another third of the mozzarella, half of the Parmesan, and another third of the sauce. Top with a third layer of noodles and the remaining third of the sauce and remaining cheeses. Bake in a preheated 350-degree oven for 40 to 45 minutes or until the cheeses are melted and the lasagne is bubbly. If possible, let the lasagne sit for a few minutes before serving. To serve, cut the lasagne into rectangles with the sharp edge of a wide spatula. Lift the individual servings from the pan with the spatula.

Makes about 9 servings.

HAMBURGER SKILLET FLORENTINE

(main dish)

1   pound lean ground beef
1   large onion, finely chopped
1   garlic clove, minced
1   15-ounce can tomato sauce
1   10-ounce package chopped frozen spinach, thawed and well drained
½   teaspoon dried oregano leaves
½   teaspoon dried basil leaves
1   bay leaf
¼   teaspoon salt
¼   teaspoon black pepper, preferably freshly ground
¼   cup grated Parmesan cheese

In a large heavy skillet over medium-high heat, cook the ground beef, onion, and garlic, breaking up the meat with a spoon, until the meat is browned and the onion is tender. Drain off and discard any excess fat. Stir in the tomato sauce, spinach, oregano, basil, bay leaf, salt, and pepper. Bring to a boil. Cover, lower the heat, and simmer for about 15 to 20 minutes, or until the flavors are blended and the spinach is cooked. Stir in the Parmesan cheese and serve.

Makes 4 to 5 servings.

Sprouts, Bean and Alfalfa

The technique of germinating beans and seeds to produce edible sprouts was discovered in China hundreds of years ago.

Nowadays, the most popular varieties for sprouting are alfalfa seeds and mung beans. Alfalfa sprouts are quite thin and about 1 to 2 inches long, with a few green leaves at the tip. They are usually eaten raw in salads or on sandwiches or as a garnish. Mung bean sprouts—the type used in Chinese cooking—are bulkier, measure about 2 to 3 inches in length, and are typically eaten before the leaves have fully developed.

Availability: Both fresh mung bean sprouts and alfalfa sprouts are often available year round in the fresh produce section of supermarkets. The alfalfa sprouts are usually packaged in square plastic containers, while the mung bean sprouts may be loose and sold by the pound. (Bean sprouts are also available canned, but hardly compare in taste and texture to the fresh ones.)

To grow sprouts at home, put about 2 tablespoons of dry mung beans or alfalfa seeds in a large, very clean, wide-mouth jar. (Empty mayonnaise jars work well, for example.) Cover the top with a few layers of cheesecloth or a piece of nylon stocking held securely in place with a strong rubber band. The covering, which will be left in place until the sprouts are ready, permits good air circulation and easy drainage. No other equipment is needed.

Rinse the seeds thoroughly with warm tap water several times. (Just run the water right through the covering, and invert the jar to drain.) Then fill the jar about two-thirds full with warm water. Soak the seeds for about 4 to 8 hours or overnight. Pour off the soaking water and rinse the seeds well. Turn the jar upside down to drain it completely, as excess moisture promotes molding. Lay the jar on its side to give the sprouts maximum area for growth.

Keep the jar on the kitchen countertop or anywhere you won't forget about it, but not in direct sunlight. Rinse and drain the seeds once or twice daily, using the technique described above, until the sprouts are grown. The seeds will increase enormously in volume as they sprout. In 3 to 5 days (depending on variety and room temperature), you'll have a jar full of succulent, ready-to-eat young vegetables. "Harvest" your "crop" when the alfalfa sprouts are 1 to 2 inches long, and the bean sprouts are 2 to 3 inches long.

Choosing the Best: When purchasing bean or alfalfa sprouts, be sure they are very fresh. The shoots should be white and firm. Avoid any sprouts that are brownish, or look limp, slimy, or withered. Root tips and any leaves should be dry, not mushy or shriveled.

Storage: Refrigerate purchased or home-grown sprouts in a loosely covered container or open plastic bag. (Don't keep them in an airtight container or bag, as this causes quick deterioration; they need to "breathe.") The sooner the sprouts are eaten, the better the taste and the higher the nutritional content. However, they will stay fresh for a few days if they are kept cool and dry.

Nutritional Value: Due to the chemical changes that occur during germination, sprouting mung beans and alfalfa seeds increases their protein content tremendously, as well as their vitamins, particularly vitamins C and B. Also, volume for volume, sprouts are much lower in calories than the beans and seeds from which they come. A half cup of cooked beans has more than 100 calories, while the same amount of sprouts has only 10 to 25, depending on the variety. Sprouts have an exceptionally high protein-to-calorie ratio.

Preparation and Basic Cooking: A quick rinse is all that sprouts need. During sprouting, the husks of the beans or seeds may loosen or fall off. They can be eaten with the sprouts or rinsed off, as desired.

Alfalfa sprouts are usually eaten raw. Bean sprouts may be eaten raw, or they can be stir-fried with a bit of salt or soy sauce. After rinsing, drain them very well on paper towels. In a wok or large skillet over medium-high heat, heat a few tablespoons of peanut or vegetable oil until it is quite hot. Add the bean sprouts and stir-fry them just until heated through and still quite crisp, about 2 minutes. Toss with salt or soy sauce to taste.

Simple Serving Suggestions: Raw alfalfa sprouts make wonderful "nests" for hard-cooked or deviled eggs, and look very attractive when small bunches are used in a composed or tossed salad. They are also good on sandwiches.

Mung bean sprouts are tasty eaten raw in tossed salads, as well as lightly stir-fried for only a minute or two as part of an Oriental-type dishes that may include other vegetables and/or meat or poultry.

BEAN SPROUT AND SPINACH SALAD

1	small onion, shredded
3½	tablespoons vegetable oil
4	cups fresh bean sprouts
3	tablespoons apple cider vinegar
2	tablespoons sugar
1	tablespoon ketchup
½	teaspoon powdered mustard
⅛	teaspoon salt
⅛	teaspoon black pepper, preferably freshly ground
5 to 6	cups lightly packed, well-washed, and torn fresh spinach leaves
5 to 6	medium-sized fresh mushrooms, sliced
5 to 6	medium-sized radishes, thinly sliced

Combine the onion and the oil in a large skillet over high heat. Cook, stirring, for 1 minute, or until the onion is limp but not brown. Add the bean sprouts and cook, stirring, for 30 seconds longer; do not overcook. Using a slotted spoon, transfer the onion and sprouts to a bowl and set aside to cool.

To prepare the dressing, add the vinegar, sugar, ketchup, mustard, salt, and pepper to the skillet and cook, stirring, until the sugar dissolves. Set the dressing mixture aside to cool.

Combine the spinach, mushrooms, and radishes in a salad bowl. When the sprouts and the dressing mixture have cooled to room temperature, toss well with the other vegetables and serve.

Makes 5 to 7 servings.

HEARTY "CALIFORNIA" SANDWICH

(light main dish)

8	slices whole wheat or multi-grain bread
	About ¼ cup mayonnaise or similar sandwich "spread"
4	leaves fresh romaine lettuce
1	large or 2 small tomatoes (preferably vine ripened), sliced

1   medium-sized ripe avocado, peeled, seeded, and sliced
6 to 8   ounces Monterey Jack (or similar) cheese, cut into 4 large slices
        About 2 cups loosely packed fresh alfalfa sprouts

If desired, toast the bread. Spread 1 side of each slice lightly with mayonnaise. Then top each of 4 slices with 1 lettuce leaf, 2 or 3 slices each of tomato and avocado, 1 slice of cheese, and about ½ cup of sprouts. Top with the remaining bread slices, mayonnaise side down. Cut each sandwich into halves or quarters and serve.

Makes 4 servings.

ORIENTAL-STYLE CHICKEN AND BEAN SPROUTS

(main dish)

Quick and easy, this makes a good last-minute dinner.

4   medium-sized chicken breast halves, skinned and boned (about 1 pound meat)
1½   tablespoons cornstarch
2   egg whites, lightly beaten
½   teaspoon salt (or to taste), divided
1½   tablespoons dry sherry
3   tablespoons peanut or vegetable oil, divided
4   cups fresh bean sprouts
2   scallions, including green tops, thinly sliced on the diagonal
1   garlic clove, minced

Cut the chicken breasts into very thin slices about 1½ to 2 inches long. (This is easiest to do when the chicken is partially frozen.) Toss the chicken slices with the cornstarch. Add the egg whites, a pinch of the salt, and the sherry and mix well.

In a wok or very large skillet over medium-high to high heat, heat 1 tablespoon of the oil until very hot. Add the chicken mixture and stir-fry for a few minutes, or until the chicken is opaque and cooked through. Use the stirring spoon to scrape up any bits that stick to the pan. Temporarily transfer the cooked chicken to a large serving dish.

Heat the remaining 2 tablespoons of oil in the pan and add the bean sprouts, scallions, and garlic. Sprinkle with the remaining salt. Stir-fry just until the sprouts are heated through but still quite crunchy, about 1 to 2 minutes. Return the chicken to the pan and stir-fry the sprouts and chicken together for about 30 seconds; then serve.

Makes about 4 servings.

ORIENTAL STIR-FRIED STEAK

(main dish)

Marinade
¼ cup soy sauce
½ cup dry sherry
2 teaspoons finely chopped gingerroot (or ½ teaspoon ground ginger)
1 garlic clove, minced
1 tablespoon packed light brown sugar
¼ teaspoon powdered mustard
3 drops Tabasco sauce
⅛ teaspoon black pepper, preferably freshly ground

Meat and Vegetables
1 pound sirloin steak, cut into 3- by ¼-inch strips
2 tablespoons vegetable oil
1 large onion, coarsely chopped
1 sweet red or green pepper, cut into 1-inch squares
1 medium-sized celery stalk, thinly sliced
3 cups small broccoli flowerets
2 cups fresh bean sprouts
2 tablespoons cold water
1 tablespoon cornstarch

To Serve
   Hot cooked white rice

In a medium-sized bowl, stir together all the marinade ingredients. Add the steak strips and stir to coat them well. Cover and refrigerate for 1½ to 2 hours.

Remove the meat from the marinade with a slotted spoon and reserve the marinade. In a large heavy skillet over medium-high heat, heat the oil. Add the meat and stir-fry for about 3 to 4 minutes, or until browned and cooked through. Add the onion, red or green pepper, and celery and stir-fry an additional 3 minutes. Stir in the reserved marinade and the broccoli. Bring to a boil. Cover and lower the heat to medium-low. Steam the vegetables and meat for about 3 to 4 minutes, or until the vegetables are crisp-tender and have absorbed the marinade flavor. Add the bean sprouts and cook, stirring, about 1 minute longer, or just until the sprouts are heated through.

Stir together the water and cornstarch in a small bowl. Add to the pan. Stir until the sauce is lightly thickened. Serve immediately over hot cooked white rice.

Makes 4 to 5 servings.

Squash, Summer

Summer squash are those types eaten while still immature and tender. At this stage of development, their skin and seeds are still soft and edible, and their flesh has a high water content. Thus, they do not store well, and must be eaten shortly after harvesting.

Before the advent of modern agricultural methods and year-round cultivation in warmer areas of the country, this type of squash was available only during the summer and early fall. Now that the growing season has been greatly extended, the term "summer squash" is something of a misnomer. Thus, summer squash are occasionally referred to as "soft-shelled squash."

Summer squash may be eaten raw and are tasty in this state, unlike their cousins, the hard-shelled winter squash. Both kinds are indigenous to the Americas.

Following are the major types of summer squash commonly available. Though each has a different skin color and shape, they are actually somewhat similar in taste, and most can be used interchangeably in recipes. All have relatively smooth skin and whitish flesh.

Cocozelle is very similar to zucchini (see below). It is cylindrical with skin that has yellow or white stripes on a dark green background.

Pattypan is also known as *cymling* or *scallop*. This disk-shaped squash is thicker in the center than at the attractively scalloped edges. When the squash is very young and small, its skin is pastel green, but becomes whiter with increased maturity.

Yellow crookneck is quite similar to Yellow Straightneck (below), but has a neck that is gracefully curved over.

Yellow straightneck resembles a small bowling pin, with a narrow neck and much wider base. Its skin color ranges from very pale yellow to darker yellow as it matures.

Zucchini is sometimes called "Italian squash" because it is used so much in Italian cuisine. It is straight and cylindrical with mottled-looking dark green skin. This is the most popular type of summer squash.

   Availability: In most areas, zucchini, and possibly also yellow squash, are available year round. The other types are available at scattered times in the year, depending on the area of the country. The peak season for all of them is May through October.

   Choosing the Best: As noted earlier, summer squash are eaten while still immature, and (unless taken to extremes) the smaller and younger, the better. Interestingly, "baby" summer squash have become the rage in some gourmet restaurants and specialty produce stores, as much for their dainty look as for their taste.
   Choose summer squash with skin that is shiny and undamaged. Bruises or soft spots can cause summer squash to deteriorate rapidly. In general, avoid large, overly mature squash with skin that is dull or beginning to harden.

   Nutritional Value: Summer squash are high in vitamin C, and have fair amounts of vitamin A, niacin, potassium, and magnesium. They are also high in fiber and extremely low in calories.

   Storage: Refrigerate summer squash in a loosely closed plastic bag or container or a vegetable crisper, and use within 2 to 3 days for best results (though some summer squash may keep a week or longer).

   Preparation and Basic Cooking: Summer squash can be readied for cooking simply by washing, then trimming off the stem end, and possibly also the blossom end. The skin and seeds are entirely edible.
   To cook summer squash, leave small ones whole, or cut larger ones into chunks, slices, or spears. Put the squash into a steamer basket over simmering water, and cover the pan tightly. Steam the squash just until it is tender when pierced with a fork; the exact time will vary depending on the maturity of squash and the size of pieces. Slices that are ½ inch thick should take only about 4 to 6 minutes.
   Whole squash and large chunks may be boiled, if preferred; however, slices and spears may get "waterlogged" using this method.

Thin slices of squash, as well as grated squash, are also very good sautéed in a little butter or oil. If grated squash is used, it should be squeezed first to remove any excess liquid. Sauté the squash just until it is lightly browned, about 2 to 5 minutes.

Simple Serving Suggestions: Sautéed squash is very good with chopped herbs, such as basil, oregano, and parsley. It is also quite nice mixed into an omelet. Raw spears or slices of zucchini or yellow squash can be used as crudités with vegetable dips, and smaller pieces may be tossed in a salad. All kinds of summer squash can be hollowed out leaving a ¼-inch-thick shell, and stuffed with a seasoned ground meat and/or rice mixture, then baked with or without a topping of tomato sauce, or possibly some grated cheese.

MARINATED ZUCCHINI AND TOMATO SALAD

Two vegetables that seem to be particularly prolific in home gardens are zucchini and tomatoes. This salad takes advantage of an abundant harvest.

5	medium-sized zucchini (about 8 ounces each)
⅓	cup good-quality olive oil
⅓	cup red wine vinegar or apple cider vinegar
2	teaspoons honey
1	small garlic clove, finely minced
1	very small onion, grated (or 2 tablespoons instant minced onions)
¼	teaspoon salt
⅛	teaspoon black pepper, preferably freshly ground
4	large vine-ripened tomatoes, cut into small wedges
1	sweet green pepper, cut into 2-inch strips
2 to 3	scallions, including green tops, thinly sliced
2	tablespoons finely chopped fresh parsley leaves
2	teaspoons dried basil leaves (or 2 tablespoons chopped fresh basil)

Cut the ends from the zucchini; then cut them crosswise into 2-inch-long sections. Steam the zucchini in a small amount of water until just crisp-tender, about 4 to 5 minutes. Cool; then cut each piece lengthwise through the center into 6 to 8 narrow, 2-inch-long wedges.

To make the marinade, combine the oil, vinegar, honey, garlic, onion, salt, and pepper in a small jar with a tight-fitting lid; then cover it and shake well.

Put the zucchini, tomatoes, green pepper, scallions, parsley, and basil in a serving bowl. Add the marinade and toss the vegetables gently. Cover and chill the salad for about 2 to 3 hours to give the flavors a chance to mingle. Toss again before serving.

Makes about 6 servings.

SQUASH AND GREEN PEPPER SKILLET

(side dish)

2   tablespoons olive oil
1   medium-sized onion, finely chopped
1   small garlic clove, minced
1   medium-sized yellow squash, thinly sliced
1   large sweet green pepper, cut into 1-inch squares
1   celery stalk, thinly sliced
¼   cup dry white wine
1   8-ounce can tomato sauce
½   teaspoon dried oregano leaves
    Pinch of black pepper, preferably freshly ground

In a large skillet over medium-high heat, combine the olive oil, onion, and garlic. Cook, stirring constantly, until the onion is tender. Add the squash, green pepper, and celery and continue to cook, stirring, for 2 minutes. Add all the remaining ingredients and combine well. Lower the heat, cover, and simmer, stirring occasionally, for about 15 minutes, or until the green pepper is tender.
Makes 3 to 4 servings.

PASTA-VEGETABLE CASSEROLE

(side dish or light main dish)

1   tablespoon olive oil
1   medium-sized onion, finely chopped
1   small garlic clove, minced
2   15-ounce cans tomato sauce
2   cups thinly sliced zucchini (about 2 medium-sized zucchini)
1   sweet red pepper, coarsely chopped (if unavailable, substitute a sweet green pepper)
1   sweet green pepper, coarsely chopped
1½ teaspoons dried oregano leaves
1½ teaspoons dried basil leaves
½   teaspoon salt
¼   teaspoon black pepper, preferably freshly ground
    Generous 2½ cups uncooked elbow macaroni
6   ounces mozzarella cheese, grated (1½ cups packed)
⅓   cup grated Parmesan cheese

CONVENTIONAL METHOD

In a small saucepan, heat the oil over medium-high heat. Add the onion and garlic and cook, stirring carefully to prevent the vegetables from burning, until the onion is soft. Add the tomato sauce, zucchini, red and green peppers, oregano, basil, salt, and black pepper. Bring to a boil. Cover, lower the heat, and simmer the sauce for about 15 minutes, or until the pepper and the zucchini are tender.

Meanwhile, cook the pasta according to the package directions. Drain well in a colander. To assemble the casserole, spread half of the pasta in the bottom of a lightly greased or nonstick spray-coated 2½-quart baking dish. Spread half of the sauce over the pasta. Top with one half each of the mozzarella and Parmesan cheeses. Repeat the layers. Bake uncovered in a preheated 350-degree oven for 20 to 25 minutes, or until heated through.

MICROWAVE METHOD

Cook the pasta according to package directions and drain well in a colander. In a 2-quart round casserole, combine the oil, onion, garlic, zucchini, and peppers. Place in the microwave oven and cook, covered, on high power for 6 to 7 minutes, turning the casserole a quarter turn after 3 minutes. The vegetables should be almost tender. Add the tomato sauce and seasonings. Cook on high power, covered, for an additional 3 minutes, or until the sauce is hot. Assemble the casserole as directed above. Cook, covered, in the microwave for about 4 to 6 minutes, or until heated through.

Makes 9 to 10 servings as a side dish or about 6 servings as a light main dish.

BROILED PATTYPAN SQUASH

(side dish)

This simple cooking method preserves the delicate taste of the squash.

  3 **pattypan squash (each about 3-inches in diameter), trimmed and cut vertically into ¼-inch-thick slices**
1½ **tablespoons melted butter or margarine**
      Salt and freshly ground black pepper to taste
      Paprika for garnish (optional)

Evenly brush the squash slices on one side with melted butter. Lay, unbuttered side down, on a foil-lined broiler pan. Lightly sprinkle the slices with salt and pepper. Place the pan about 3 inches from the heating element in a preheated broiler. Broil for 4 to 5 minutes, or until the slices are lightly browned. Turn the slices over and evenly brush the second side with butter. Sprinkle with salt and pepper. Also sprinkle with a bit of paprika, if desired. Return the pan to the broiler and broil the slices for about 3 to 4 minutes longer, or until they are lightly browned.

Makes 4 to 6 servings.

MEDITERRANEAN SAUTÉ

(side dish)

1	tablespoon olive or vegetable oil
½	tablespoon butter or margarine
2	garlic cloves, minced
1	small onion, finely chopped
2	medium-sized zucchini, cut into 1½-inch-long and ¼-inch-thick sticks
1	medium-sized yellow squash, cut into 1½-inch-long and ¼-inch thick sticks
1	small sweet red pepper, cut into 1½-inch-long and ¼-inch-wide strips (if unavailable, substitute 1 small sweet green pepper)
1	small tomato, peeled and chopped
¼	cup finely chopped fresh parsley leaves
⅛	teaspoon dried basil leaves
	Generous ⅛ teaspoon salt
⅛	teaspoon black pepper, preferably freshly ground
1 to 2	drops Tabasco sauce (optional)

Combine the oil and butter in a large skillet or sauté pan over medium-high heat. Add the garlic and onion and cook, stirring, for 1 minute. Stir in the zucchini and yellow squash and red pepper, and cook, stirring constantly, for 2 minutes. Add the remaining ingredients and cook about 1½ minutes longer, or until the squash are just tender and tomatoes are heated through. Remove from the heat and serve.

Makes about 4 servings.

ZUCCHINI-CHEESE CASSEROLE

(side dish or light main dish)

This dish is similar to a crustless quiche. If desired, yellow or pattypan squash may be substituted for the zucchini.

1   **pound zucchini, ends trimmed**
3   **large eggs, lightly beaten**
½   **cup part-skim ricotta cheese (if unavailable, substitute regular ricotta or small curd cottage cheese)**
1   **cup fresh whole wheat or white bread crumbs (made in a food processor or blender)**
2   **tablespoons grated or finely minced onion**
3   **tablespoons finely chopped fresh parsley leaves**
¼   **teaspoon dried marjoram leaves**
¼   **teaspoon dried basil leaves**
¼   **teaspoon salt (or to taste)**
⅛   **teaspoon black pepper, preferably freshly ground**
4   **ounces Swiss, mozzarella, or mild Cheddar cheese, grated (1 cup packed), divided**

Coarsely grate the squash with a grater or a food processor. Let it rest for about 5 minutes; then squeeze out any excess liquid.

Meanwhile, in a medium-sized bowl, combine the remaining ingredients, *except* 2 tablespoons of the grated cheese, and mix well. Stir in the squash. Pour the mixture into a greased or nonstick spray-coated, shallow 1½-quart casserole or 10-inch-diameter quiche pan (or equivalent). Sprinkle the reserved cheese on top. Bake, uncovered, in a preheated 350-degree oven for about 25 to 30 minutes, or until the top is firm and lightly browned.

Makes about 6 side-dish servings or about 3 light main-dish servings.

ZUCCHINI STUFFED WITH TUNA AND CHEESE

(main dish)

3   medium-sized zucchini (each about ½ pound and about 7 inches long)
1   16-ounce can tomatoes, including juice
2   tablespoons olive or vegetable oil
1   medium-sized onion, finely chopped
1   garlic clove, minced
1   6½-ounce can water-packed tuna, drained and flaked
½   teaspoon dried oregano leaves
½   teaspoon dried basil leaves
⅛   teaspoon black pepper, preferably freshly ground
4   ounces mozzarella cheese, preferably part-skim, grated (1 cup packed)
2   tablespoons grated Parmesan cheese

Trim off the stem and blossom ends of each zucchini, removing as little as possible. Cut each zucchini in half lengthwise. Use a metal melon baller, a serrated grapefruit spoon, or a small knife to carefully scoop out the flesh in the center of each half, leaving a sturdy ¼-inch shell. Finely chop up the scooped-out flesh and set it aside.

Drain the juice from the can of tomatoes into a measuring cup. There should be at least ½ cup; add water if necessary. Pour the juice into a greased or nonstick spray-coated 12- by 7-inch (or equivalent) baking dish. Arrange the shells in the sauce, skin side down.

In a large skillet over medium-high heat, heat the oil; then cook the onion and garlic until they are very lightly browned. Add the reserved chopped zucchini and cook, stirring often, until it is lightly browned. Chop the drained canned tomatoes and add them to the skillet with the tuna, oregano, basil, and pepper. Cook, stirring often, until most of the liquid has evaporated from the skillet. Remove the skillet from the heat. Mix together the mozzarella and Parmesan, and stir about two thirds of the cheese into the tuna mixture.

Spoon the tuna mixture into the shells, heaping it a bit if necessary. Sprinkle the remaining cheese on top. Bake the casserole in a preheated 375-degree oven for about 30 minutes, or until the zucchini shells are tender and the cheese is lightly browned.

Makes about 4 servings.

GROUND BEEF AND SQUASH SKILLET

(main dish)

1   pound lean ground beef
1   medium-sized onion, finely chopped
1   garlic clove, minced
2   celery stalks, thinly sliced
1   cup beef broth or bouillon (reconstituted from cubes or granules)
¾   teaspoon dried marjoram leaves
¾   teaspoon dried basil leaves
¼   teaspoon dried thyme leaves
1   bay leaf
¼   teaspoon salt
⅛   teaspoon black pepper, preferably freshly ground
1   large yellow straightneck or crookneck squash, ends trimmed, cut into
    ¼-inch-thick slices (about 2 cups)
½   cup plain lowfat yogurt

To Serve
    Hot cooked white or brown rice

In a large skillet over medium-high heat, cook the ground beef, onion, garlic, and celery, breaking up the meat with a spoon, until the meat is brown and the onion is soft. Drain off and discard any excess fat. Add the broth, marjoram, basil, thyme, bay leaf, salt, and pepper. Stir to mix well.

Arrange the squash on top of the meat mixture. Lower the heat, cover, and simmer for about 25 to 30 minutes, or until the squash is very tender. Stir the squash into the meat mixture.

Remove the skillet from the heat. Add the yogurt and stir to combine well. Turn the heat to very low and return the skillet to the heat. Cook, being careful not to boil the mixture, for about 7 or 8 minutes so that the flavors can blend. Serve over hot cooked rice.

Makes 4 to 5 servings.

ZUCCHINI-NUT QUICK BREAD

Zucchini, walnuts, and spices team up to lend this easy bread a nutty-sweet flavor.

2½ cups enriched all-purpose or unbleached white flour
1¼ cups whole wheat flour
1 cup packed light brown sugar
1 tablespoon baking powder
½ teaspoon baking soda
1¼ teaspoons ground cinnamon
¾ teaspoon ground mace
½ cup lowfat or skim milk
1 large egg
3 tablespoons vegetable oil
½ teaspoon vanilla extract
1 large (unpeeled) zucchini (about ¾ pound), finely shredded or grated and well drained
½ cup finely chopped walnuts

Thoroughly stir together the flours, brown sugar, baking powder, baking soda, cinnamon, and mace in a large bowl.

In a small bowl, beat together the milk, egg, oil, and vanilla with a fork until well blended and smooth.

Stir the zucchini and walnuts into the dry ingredients until distributed throughout. Add the liquid ingredients to the dry mixture, gently stirring until the batter is thoroughly blended but not overmixed. Spoon the batter into a well-greased 9- by 5- by 3-inch (or similar 2-quart) loaf pan, smoothing the top surface and spreading it out to the edges.

Bake the loaf in a preheated 375-degree oven for 15 minutes. Lower the oven temperature to 350 degrees and continue baking for 50 to 55 minutes longer, or until the top is nicely browned and a toothpick inserted in thickest part of the loaf comes out clean. Remove the pan to a wire rack and cool for 10 minutes. Then remove the loaf from the pan and let stand on the rack until completely cooled. Serve the bread at room temperature or slightly chilled.

Makes 9 to 11 servings.

Squash, Winter

Unlike summer squash, winter squash are eaten only after they have fully matured and developed tough skins and large seeds. In fact, they have been specially adapted for winter storage; thus, their name. Occasionally, they are referred to as "hard-shelled squash." Although there has traditionally been a distinction between "pumpkins" and "squashes," a pumpkin is actually just one type of winter squash.

Squash are native to the New World, and were discovered here by the early colonists. Squash seeds were then taken back to Europe, where the vegetable quickly became popular.

Winter squash are characteristically brightly colored, with shells of brilliant orange, yellow, green, and white—sometimes solid and sometimes splashed with a gay mixture. Indeed, some types are grown mainly for ornamental purposes. Others are popularly used for both cooking and decoration. The pumpkin, which may be transformed into a pie, cake, or even a whimsical jack o' lantern, is among this latter group (though different varieties of pumpkin are best for each use). Most winter squash have tender, sweet flesh in colors ranging from pale yellow to dark orange, depending on the variety.

Though the raw seeds of winter squash are inedible, they can be eaten if they are toasted. The crunchy seeds are usually hulled with the fingers or teeth, and make an enjoyable, healthful snack. (Interestingly, a few varieties of pumpkin are now being grown specifically for their unique, and convenient, hull-less seeds!)

Some of the most popular types of winter squash include the following:

Acorn, which is occasionally called "Table Queen," is deeply ribbed and acorn-shaped. This small squash (usually about 1 to 2 pounds) has dark green skin, sometimes highlighted with a touch of orange. (The "Golden Acorn" squash has solid golden-orange skin, and looks almost like an odd-shaped, miniature pumpkin. "Golden Nugget" is similar, but slightly smaller.) The moist flesh is dark yellow to pale orange. There is a large seed cavity in the center of the squash.

Buttercup is similar in size to the acorn, but has a turban-like cap at its blossom end. Thus, it is sometimes referred to by the more generic term "turban squash." It has dark green skin with grayish areas, and dark yellow to orange flesh. (There are other, larger, more brightly colored, turban squash; however, they are grown primarily for ornamental purposes.)

Butternut is long and cylindrical at the top, with a broad, bulbous base, which contains the seed cavity. The cream-colored to pale brown skin is relatively thin and smooth, making it one of the best choices when peeling is required before cooking. The fine-grained flesh is a brilliant orange.

Hubbard is one of the larger popular winter squashes. It may be pear-shaped, or it may have a globular middle with a narrow, neck-like protrusion at the stem end and a very short one at the blossom end. The skin is very bumpy and ridged, and may range in color from dark green to blue-gray to orange. Its sweet dark yellow to orange flesh is drier than some other squash, making it ideal for baked goods.

Large banana is one of the largest winter squashes—up to 1½ to 2 feet in length. It is nearly cylindrical, tapering slightly at both ends, with pale gray to pinkish-orange skin that is wrinkled and bumpy. The finely textured flesh is creamy-orange in color. (Due to its size, this squash is sometimes sold cut into pieces.)

Pie or sugar pumpkin is a variety of pumpkin grown especially for cooking. It is smaller, and has sweeter, denser, finer-grained flesh than the larger pumpkins grown for ornamental purposes. (Note: In some other countries, the term "pumpkin" may be used to describe a different group of squash than in the United States.)

Spaghetti squash is oblong in shape, with pale to dark yellow skin that intensifies in color as the squash matures. The flesh is pale yellow, and separates into long, thin, pasta-like strands when cooked, giving this squash its name. In fact, the strands can be substituted for cooked spaghetti in some recipes.

Availability: Many types of winter squash are available year round. The peak is from October through February. Those winter squash sold during the spring and early summer have sometimes been stored all winter, and may not be top quality.

Choosing the Best: Choose fully mature winter squash that are heavy for their size. They should have firm, intensely colored skin that shows no sign of decay, and does not give easily to pressure. (In general, as winter squash matures, its skin hardens and darkens in color.) Avoid winter squash with any cuts or blemishes, as they will not store as well. If possible, buy whole squash rather than pieces, for the same reason.

Nutritional Value: The orange- and golden-fleshed winter squash far outshine their summer cousins when it comes to providing generous amounts of beta-carotene, a vitamin precursor that converts to vitamin A in the body. (Beta carotene may be helpful in preventing lung cancer.) Winter squash are also a moderate source of Vitamin C, as well as some minerals such as potassium and magnesium.

Storage: Most winter squash will keep at room temperature for at least 1 week, and several months in a cellar or other dry, cool storage area. In the refrigerator, they will keep about 1 to 4 weeks or longer. Cut squash should be used as soon as possible.

Preparation and Basic Cooking: Small squash may be cooked whole (pierced deeply with a small knife in a few places to release steam) or cut in half; larger ones cook more evenly if first cut into pieces. If squash are cut, the seeds and stringy fibers should be removed from the seed cavity before cooking. (The seeds may be roasted and eaten, if desired; see below.) When squash are cooked whole, the seeds are removed immediately after cooking.

(Spaghetti squash should be cooked whole to best produce the characteristic long strands. See the recipe for Spaghetti Squash with Herbs and Cheese on page 259 for more explicit directions.)

Most winter squash can be baked in a 350- to 375-degree oven for about 40 to 80 minutes, depending on the thickness of the flesh, the size of the pieces, and the variety. Those with denser, finer-grained flesh take longer than the others.

Small squash can be baked whole in a microwave oven. They should be pierced, and rotated occasionally for even cooking. (A single acorn squash that weighs about 1¼ pounds will take about 8 minutes on full power; 2 will take about 12 to 14 minutes. Let the squash rest about 5 minutes after cooking, to complete the internal cooking. When the flesh is tender, the squash skin will "give" to gentle pressure.)

When baking squash pieces in a conventional oven, put the squash cut side down in a greased baking pan, or leave it cut side up and brush the cut edges and seed cavity with melted butter. If desired, cover the pan loosely with foil to keep in moistness. Also, ½ inch of water may be put in the bottom of the pan.

Squash can also be peeled and cut into chunks or cubes and steamed, braised, simmered, sautéed, or baked. Slices that are about ½ inch thick should be boiled for about 7 to 9 minutes, or steamed a few minutes longer, or until tender.

To toast raw squash seeds, first wash them thoroughly, rubbing off the stringy fibers. Drain the seeds well. Toss the damp seeds with a small amount of salt, if desired, and spread them on a lightly greased or nonstick spray-coated baking sheet. Bake in a 350-degree oven for about 20 minutes, stirring them occasionally, until they are golden.

Simple Serving Suggestions: Serve baked squash halves or large pieces right in the shell, topped with a little butter and perhaps a touch of maple syrup or molasses. Or, scoop cooked squash from the shell, mash it, and stir in spices, such as cinnamon, nutmeg, and allspice, some chopped nuts, and a bit of brown sugar. Plain mashed squash can be used in many baked goods and pies, particularly in recipes calling for canned pumpkin. (If the mashed squash seems to be very wet and loose, heat it in a saucepan over medium heat, while stirring, to evaporate some moisture.) Leftover mashed squash (including canned pumpkin) may be stored in an airtight container in the refrigerator for 3 to 4 days, or frozen for up to 3 months or longer.

BUTTERNUT SQUASH SOUP

1   2½-pound butternut squash
1   tablespoon butter or margarine
1   small garlic clove, minced
1   large onion, chopped
5   cups beef stock, broth, or bouillon (reconstituted from cubes or granules)
1   large bay leaf
¼   teaspoon black pepper, preferably freshly ground
⅛   teaspoon ground allspice
     Pinch of dried thyme leaves
     Pinch of cayenne pepper
1   large tomato, peeled, seeded, and coarsely chopped
½   teaspoon salt (reduce to ¼ teaspoon or omit if commercial beef broth is used)

Croutons
1½   tablespoons butter or margarine
1½   cups ½-inch slightly stale French or Italian bread cubes

Trim off the stem end and remove the peel from the squash with a very sharp paring knife or vegetable peeler. Cut the squash in half lengthwise with a sharp knife. Using a fork or spoon, scrape out and discard the seeds and fibers from the seed cavity. Cut the squash into 1-inch chunks.

Melt the butter in a 4- to 5-quart pot over medium-high heat. Add the garlic and onion and cook, stirring, for 4 to 5 minutes, or until the onion is limp but not browned. Stir in the squash, broth, bay leaf, pepper, allspice, thyme, and cayenne. Bring the mixture to a boil over high heat. Lower the heat, cover the pot, and simmer the mixture for 15 minutes. Stir in the tomato and salt (if used) and cook for 5 to 8 minutes longer, or until the squash is just tender.

Meanwhile, prepare the croutons. Melt the butter in a medium-sized skillet over medium-high heat. Add the bread cubes and cook, stirring, for about 4 to 5 minutes, or until they are golden brown and crisp. Set aside.

When the squash is just tender, transfer the mixture in batches to a blender and blend on medium speed until completely puréed. Return the puréed mixture to the pot and heat until piping hot. Ladle the soup into individual bowls and garnish each serving with a tablespoon or two of the toasted croutons; serve immediately.

Makes 6 to 8 servings.

LEMON-GLAZED BUTTERNUT SQUASH CUBES

(side dish)

1   **very large or 2 small butternut squash (about 3 pounds total weight)**
3   **tablespoons butter or margarine**
3   **tablespoons packed dark brown sugar**
2   **tablespoons lemon juice, preferably fresh**

CONVENTIONAL METHOD
Peel, halve, and seed the squash; then cut it into ½-inch cubes. This should yield about 6 cups of cubes, but the exact quantity is not critical.

Put the butter in a greased or nonstick spray-coated, shallow casserole or baking pan and set the pan in a preheated 400-degree oven just until the butter melts. Remove the pan from the oven and stir the brown sugar and lemon juice into the melted butter until well mixed. Stir in the squash cubes so they are all coated with the butter mixture. Cover the pan loosely with a lid or foil.

Bake the squash in the 400-degree oven for 20 minutes. Remove the foil to allow excess moisture to evaporate, and bake the squash for an additional 10 to 15 minutes, or until it is quite tender, stirring occasionally to evenly distribute the glaze.

MICROWAVE METHOD
Melt the butter in a shallow round casserole in the microwave oven. Add the brown sugar and lemon juice and mix well. Stir in the squash cubes so they are all coated with the glaze. Cover the casserole with wax paper or the casserole lid. Cook the squash on high powder, stirring often, for about 10 minutes, or until it is quite tender.

Makes about 6 servings.

EASY MICROWAVE SQUASH PUDDING

(side dish)

The microwave oven and frozen squash purée team up to make this dish quick and easy.

1 tablespoon cornstarch
1 12-ounce package frozen cooked winter squash purée, thawed
   Generous ¼ teaspoon salt
   Scant ¼ teaspoon black pepper, preferably freshly ground
1 small onion, finely chopped
1 large egg, beaten
¼ cup whole or lowfat milk
2 tablespoons butter or margarine, melted
2 ounces sharp Cheddar cheese, grated (½ cup packed)

In a medium-sized bowl, stir the cornstarch into the squash and mix well. Stir in the salt, pepper, and onion. Add the egg, milk, and butter and stir to mix well. Blend in the cheese. Transfer the squash mixture to a round, 1-quart casserole with a lid. Microwave on high power for 11 to 14 minutes, or until the squash mixture has set. Turn the casserole one quarter turn every 5 minutes during the cooking period.
   Makes 4 to 5 servings.

BUTTERNUT SQUASH WITH ONIONS AND APPLES

(side dish)

1 tablespoon butter or margarine
1 medium-sized onion, finely chopped
1 medium-sized butternut squash (about 2¼ pounds), peeled, seeded, and
   cut into ¾-inch cubes
½ cup apple cider, apple juice, or orange juice
1 medium-sized apple, peeled, cored, and diced

In a large skillet, over medium-high heat, melt the butter; then cook the onion until it is tender but not browned. Add the squash cubes and cook, stirring, for about 2 minutes longer. Pour in the cider and stir so that the squash is coated. Cover the skillet tightly, lower the heat, and simmer the squash for 10 minutes. Stir in the diced apple and simmer, covered, stirring occasionally, for about 10 minutes longer, or until the squash has been cooked to the desired tenderness.
   Makes 4 to 6 servings.

NUTTY ACORN SQUASH

(side dish)

This can be prepared in a conventional oven or in a microwave oven, which is much quicker.

3   small acorn squash (about 1 pound each)

Filling
2½   tablespoons packed dark brown sugar
2   tablespoons enriched all-purpose or unbleached white flour or whole wheat flour
½   teaspoon ground cinnamon (or a little more, to taste)
¼   teaspoon ground nutmeg
⅓ to ½   cup finely chopped walnuts or pecans (to taste)
2   tablespoons softened butter or margarine

CONVENTIONAL METHOD
Cut each acorn squash in half lengthwise; then scrape out and discard the seeds and fibers.

Place the squash halves, cut side up, in a baking dish or pan and add about ¼ inch of water to the bottom of the dish. Cover the dish loosely with aluminum foil. Bake the squash in a preheated 375-degree oven for 30 minutes.

Meanwhile, make the filling. Combine the brown sugar, flour, cinnamon, nutmeg, and nuts in a small bowl. Cut in the butter with your fingertips, a pastry blender, or two knives, until the mixture is crumbly and completely mixed. Remove the partially baked squash from the oven, and sprinkle the nut filling into the cavities, dividing it evenly. Cover the squash and continue baking for about 20 to 30 minutes longer, or until the flesh is quite tender and very easily pierced with a fork.

Serve each squash half, intact, as 1 serving. Diners can spoon the filling and squash flesh from the shell.

MICROWAVE METHOD
Halve and seed the squash as above. Arrange the halves, cut side up, in a circle on a microwave-proof platter and cover it loosely with wax paper or heavy-duty plastic wrap. Microwave the squash on high power for about 10 to 12 minutes, or until the squash flesh is almost tender. For even cooking, rotate the dish a quarter turn about every 3 minutes during the cooking period. Prepare the nut filling (as above) and sprinkle it into the cavities. Cover the squash. Microwave for about 3 to 5 minutes longer, or until the squash flesh is tender when pierced with a fork. Let the squash stand, covered, for about 5 minutes longer to complete the internal cooking.

Makes 6 servings.

SPANISH-STYLE WHITE BEANS AND SQUASH

(side dish)

- 1½ tablespoons butter or margarine
- 1 medium-sized onion, finely chopped
- 2 garlic cloves, minced
- 1 medium-sized butternut squash (1½ to 2 pounds), peeled, seeded, and cut into ½-inch cubes
- ¼ cup water (or more if needed during cooking)
- 2 15- to 16-ounce cans white beans, drained (OR 3 cups cooked and drained navy, pea, or Great Northern beans)
- 1 16-ounce can tomatoes, including juice, chopped
- ¼ teaspoon dried basil leaves
- ¼ teaspoon dried thyme leaves
- ⅛ teaspoon dried oregano leaves
- ⅛ teaspoon black pepper, preferably freshly ground

In a large skillet over medium-high heat, melt the butter; then sauté the onion and garlic until they are tender but not browned. Stir in the squash; then add the ¼ cup water and cover the skillet tightly. Lower the heat and simmer the squash for about 10 to 15 minutes, or until it is just tender. (If the mixture becomes very dry, add a few tablespoons of additional water.)

Remove the cover from the skillet and stir in the beans, tomatoes and their juice, herbs, and pepper. Simmer, uncovered, stirring often, for about 5 minutes longer, or until most of the liquid has been absorbed and the squash is coarsely mashed.

Makes about 8 servings.

SPAGHETTI SQUASH WITH HERBS AND CHEESE

(side dish)

When spaghetti squash is correctly cooked, it looks something like the pasta for which it is named; yet, it is much lower in calories and higher in nutrients. It not only makes an easy and tasty side dish, as in the following recipe, but it can also be served with meatballs or meaty tomato sauce for a main course.

1   large spaghetti squash (about 4 pounds)*
2   tablespoons butter or margarine, softened
¼   cup grated Parmesan cheese
2   tablespoons finely chopped fresh parsley leaves
     About 1 teaspoon dried basil leaves (or 1 tablespoon finely chopped fresh basil), or to taste
⅛   teaspoon black pepper, preferably freshly ground

For firm-textured, "pasta-like" strands, it is best to cook spaghetti squash whole. First, pierce the top surface of the squash deeply 3 or 4 times with a small knife, so steam can escape. Then, either bake it in a 350-degree oven for about 1½ hours, or put it in a large pot, and nearly cover it with water; then simmer it, covered, for about 45 to 55 minutes. (Alternately, cook the pierced squash in a microwave oven for about 15 to 20 minutes, rotating it often; then rest it for about 5 to 10 minutes to complete the internal cooking.) In all cases, the spaghetti squash is done when the skin gives to gentle pressure.

Split the cooked squash in half lengthwise and carefully scoop out and discard the seeds and fibers in the center. Then use the tines of a fork to gently pull out and separate the spaghetti-like strands. (Note: If the squash is undercooked, the strands will not come out easily. If it is overcooked, the strands will be mushy.)

Put the spaghetti squash strands into a large bowl and immediately toss them with the remaining ingredients. Serve hot.

Makes about 6 servings.

*Two 2-pound spaghetti squash may be substituted for the 4-pound one. Baking and boiling time will be reduced significantly. Microwave cooking time will be only slightly reduced, if at all.

PUMPKIN MUFFINS

¾ cup enriched all-purpose or unbleached white flour
¾ cup whole wheat flour
¼ cup instant nonfat dry milk powder
1 teaspoon baking powder
1 teaspoon baking soda
¾ teaspoon ground ginger
¼ teaspoon ground cloves
¼ teaspoon ground cinnamon
⅛ teaspoon salt
3 tablespoons vegetable oil
1 large egg, slightly beaten
¼ cup light molasses
3 tablespoons sugar
¼ cup water
⅔ cup canned solid-pack pumpkin (not pumpkin pie filling)
¾ cup dark raisins

In a medium-sized bowl, combine the flours, milk powder, baking powder, baking soda, ginger, cloves, cinnamon, and salt. Stir to mix well. Set aside.

In a small bowl, combine all of the remaining ingredients, *except* the raisins, and stir to mix well.

Add the pumpkin mixture to the dry mixture. Stir just until mixed. Stir in the raisins. Spoon the batter into 12 nonstick spray-coated or lightly greased muffin cups. Bake in a preheated 400-degree oven for 15 to 17 minutes, or until the muffins are puffed and lightly browned.

Makes 12 muffins.

PUMPKIN COOKIES

Here are cookies that are not only spicy and delicious but also high in beta-carotene, which the body converts to vitamin A.

½   cup butter or margarine
⅔   cup packed light brown sugar
1   large egg
¾   cup canned solid-pack pumpkin (not pumpkin pie filling)
¾   cup enriched all-purpose or unbleached white flour
¾   cup whole wheat flour
¾   teaspoon ground cinnamon
½   teaspoon baking powder
¼   teaspoon baking soda
¼   teaspoon ground cloves
¼   teaspoon ground ginger
⅛   teaspoon salt
1   cup dark raisins

In a medium-sized bowl, cream the butter and brown sugar with an electric mixer on medium speed until smooth. Beat in the egg until blended. Add the pumpkin and continue mixing until well blended.

Stir the dry ingredients together. Gradually beat them into the creamed mixture. Then stir in the raisins with a large spoon. Drop the batter by tablespoonfuls about 2 inches apart onto greased baking sheets. Bake in a preheated 375-degree oven for 10 to 13 minutes, or until the cookies are puffed and lightly browned at the edges. Transfer them to a wire rack to cool.

Makes 35 to 40 cookies.

Sweet Potatoes

$\mathbf{A}$ member of the morning glory family, the sweet potato was well ensconced in the New World before making its way to Europe. The Incas and the Mayans cultivated several types—some for food and some as an artist's coloring material. Columbus and his men were offered boiled sweet potatoes by the West Indians. And de Soto found them along the lower Mississippi River.

Sweet potatoes quickly became an important crop for the American colonists. During the Revolution and the War Between the States, when other food was in short supply, they were a dietary staple.

The sweet potato needs a warm, moist climate. Today it is grown commercially in the United States chiefly in the South and Mid-Atlantic regions.

There are literally hundreds of sweet potato varieties. Most have yellow-brown or yellow-red skins and yellow, bright orange, or yellow-red flesh, although one type could be mistaken for a white potato. Most, but not all, taste sweet. Sweet potatoes range in shape from long and slender to round.

Two main types are of commercial importance in the United States. These are commonly referred to as "dry fleshed" and "moist" although the former actually have a higher water content than the latter. The "dry flesh" variety is characteristically firm and dry when cooked, with a pale yellow-colored interior. The "moist" type is softer and sweeter, and has orange-colored flesh. Particularly in the South, these latter are sometimes referred to as "yams" although the true yam is not related to the sweet potato at all.

Availability: Sweet potatoes are available all year round, although the peak season is September through March.

Choosing the Best: Select firm sweet potatoes with smooth, uniformly colored skin. Those with knobs and deep indentations will be hard to cut without waste. Sweet potatoes are more perishable than white potatoes and decay easily. Avoid any with bruises, shriveled or discolored ends, or sunken, discolored areas on the sides. Also check for cuts, holes, or worm or grub damage.

Nutritional Value: Sweet potatoes are an excellent source of both vitamin C and vitamin A.

Storage: Sweet potatoes should be stored in a cool, dry area with good ventilation. They keep better at room temperature than in the refrigerator. Sweet potatoes will hold for 2 to 3 weeks. Discard those with any signs of decay or damage. Even if the bad part is cut away, the rest of the potato may not taste good.

Basic Preparation and Cooking: Although sweet potatoes are in no way related to white potatoes, most cooking methods suitable for white potatoes also work well with sweet potatoes. Before cooking, scrub them well and cut off any bruised or woody places. Whole sweet potatoes can be baked in a 400-degree oven for 40 minutes to 1 hour, depending on size. Before baking the sweet potatoes, pierce them with a fork so that steam can escape during cooking. Because sweet potatoes often have tiny breaks in the skin, it's best to put them in a shallow baking dish or pan. Otherwise juice may drip and burn on the oven bottom.

Boiling time for whole sweet potatoes is 20 to 30 minutes. Test for doneness with a fork. It should easily puncture the sweet potato flesh.

Sweet potatoes can also be peeled and cut into slices or cubes for quick cooking. Simmer in a covered pot with about an inch of water. Slices will need to cook about 12 to 20 minutes. Cubes will cook in 10 to 15 minutes. They should be fork-tender when done.

In general, the dry fleshed varieties require a bit longer cooking time than the moist varieties.

Sweet potatoes cook particularly well in the microwave oven. Pierce with a fork to allow steam to escape. Microwave 1 sweet potato at full power for about 5 to 8 minutes. To microwave more than 1 sweet potato, arrange them on a dish with one end pointing toward the center and the other end pointing toward the rim, like the petals of a daisy. For 2 potatoes, microwave for 8 to 10 minutes; for 3 potatoes, 10 to 13 minutes; for 4 potatoes, 13 to 17 minutes. Turn the sweet potatoes over, and rotate the plate a quarter turn once or twice during cooking. Sweet potatoes may still feel slightly firm at the end of the cooking period. Let them sit for a few minutes on the counter to become soft.

Simple Serving Suggestions: Baked sweet potatoes are excellent with butter, salt, and pepper. They can also be seasoned with a little ginger, cinnamon, or nutmeg—or whipped with butter and a bit of brown sugar.

GINGERED SWEET POTATO CASSEROLE

(side dish)

This casserole is quick and easy, yet it tastes great.

2   **pounds (about 3 large) uncooked sweet potatoes, peeled and cut into ¼-inch-thick slices (about 6 cups)**
⅔   **cup orange juice**
3   **tablespoons packed light brown sugar**
¼   **teaspoon ground ginger**
    Pinch of black pepper, preferably freshly ground
3   **tablespoons butter, cut into small pieces**

Just barely cover the sweet potatoes with water in a medium-sized saucepan. Bring to a boil over high heat. Cover, lower the heat, and simmer for about 15 minutes, or until the sweet potatoes are tender.

Drain the sweet potatoes well in a colander. Transfer them to a nonstick spray-coated or lightly greased 1½-quart casserole. Add all the remaining ingredients, *except* the butter, and stir to coat the sweet potatoes, being careful not to break them up. Dot the sweet potatoes with the butter. Bake, covered, in a preheated 375-degree oven for 20 to 25 minutes, or until the flavors have blended and the casserole is hot. Gently stir once or twice during cooking.

Makes 6 to 7 servings.

SWEET POTATO-APPLESAUCE CASSEROLE

(side dish)

For a fancier version of this dish, featuring stuffed sweet potato shells, see the variation below.

7 to 8	medium-sized sweet potatoes
4	tablespoons butter or margarine, divided
1½	cups applesauce
3 to 4	tablespoons packed dark brown sugar (to taste)
1 to 1½	teaspoons ground cinnamon (to taste)
¼	cup finely chopped walnuts or pecans

Peel the sweet potatoes; then cut them into ½-inch-thick slices or chunks. Put the sweet potatoes in a large saucepan with about 1 inch of water. Bring to a boil over high heat. Cover the pan tightly, lower the heat, and steam the sweet potatoes until they are very tender, about 25 minutes.

Drain the sweet potatoes well; then coarsely mash them with a fork or potato masher. Mix in 3 tablespoons of the butter until it is completely melted. Then stir in the applesauce. Stir in the brown sugar and cinnamon, adjusting the amounts to suit your own taste.

Transfer the sweet potato mixture to a greased or nonstick spray-coated 1½- to 2-quart casserole. Sprinkle the nuts on top and dot with the remaining 1 tablespoon of butter. (The recipe may be prepared in advance to this point.)

To complete the preparation, bake the casserole in a preheated 350-degree oven for about 35 to 45 minutes, or until it is completely heated through.

Makes about 8 servings.

Variation

TWICE-BAKED STUFFED SWEET POTATOES

Use 4 very large, evenly shaped sweet potatoes. Scrub the skins well; then prick each potato in a few places. Bake the potatoes in a preheated 400-degree oven for about 60 to 70 minutes, or until they are tender. (Or bake them in a microwave oven on high power for about 20 to 25 minutes, rotating them often, until almost tender; then let them rest for about 10 minutes to complete the internal baking.)

Cut each potato in half and scoop out the pulp leaving a ⅛-inch-thick shell. Put the shells in a large baking dish. Prepare the filling as for the casserole above; then stuff the mixture in the shells, dividing it evenly and heaping it in the center. Sprinkle the tops with the nuts and dot with the butter, as directed above. Before serving, bake the stuffed shells in a preheated 350-degree oven about 30 minutes, or until the stuffing is heated through.

Makes about 8 servings.

MICROWAVE SWEET POTATO AND FRUIT CASSEROLE

(side dish)

Prepare this tasty dish in almost no time at all in the microwave oven.

1½ cups tightly packed, coarsely chopped mixed dried fruits, such as pears, apricots, prunes, apples, and raisins
¼ cup orange juice
½ cup unsweetened apple juice
⅛ teaspoon salt
¼ teaspoon ground cinnamon (optional)
3 hot, cooked, medium-sized sweet potatoes (about 1½ pounds), peeled and cut into 1-inch cubes
2 tablespoons butter or margarine

In a 1-quart casserole, combine the fruit, juice, salt, and cinnamon (if used). Cover and microwave on full power for 4 to 5 minutes, or until the fruit is moist and tender. Give the casserole a quarter turn and stir once during the cooking period. Stir in the sweet potatoes, being careful not to break them up. Dot the sweet potatoes and fruit with the butter. Cover and microwave on full power for 2 to 3 minutes longer, giving the casserole a quarter turn and stirring once during cooking. Let the casserole stand for about 5 minutes after cooking to allow any excess liquid to be absorbed.

Makes 5 to 6 servings.

SOUTH SEAS DINNER WITH MEATBALLS

(main dish)

Meatballs
1 pound lean ground beef
1 tablespoon instant minced onions
1 large egg
½ cup plain cracker crumbs or dry bread crumbs
¼ teaspoon powdered mustard
¼ cup ketchup
½ teaspoon salt
¼ teaspoon black pepper, preferably freshly ground

Sweet Potato and Sauce

1   tablespoon butter or margarine
1   large onion, finely chopped
1   8-ounce can juice-packed pineapple chunks, including juice
⅔   cup orange juice
¼   cup dark raisins
1   tablespoon soy sauce
½   teaspoon apple cider vinegar
2   teaspoons packed light brown sugar
⅛   teaspoon black pepper, preferably freshly ground
1   large sweet potato, cooked, peeled, and cut into 1-inch cubes

To Serve

Hot cooked white or brown rice

Combine all the meatball ingredients in a medium-sized bowl and mix them together well. Form the meat mixture into about 24 balls, using 1 generous tablespoon of the mixture for each. Place the meatballs in a shallow baking pan and bake in a preheated 350-degree oven for 12 to 15 minutes, or until they are nicely browned.

While the meatballs are cooking, melt the butter in a large saucepan over medium-high heat. Add the onion and cook, stirring constantly, until it is soft. Add the pineapple and its juice, orange juice, raisins, soy sauce, vinegar, brown sugar, and pepper. Stir to mix well. Simmer the sauce, covered, over low heat for about 10 minutes.

When the meatballs are ready, remove them from the baking sheet with a slotted spoon. Add them to the sauce, along with the sweet potato. Stir the mixture to coat the meatballs and the sweet potato cubes with the sauce, but be careful not to break them up.

Cover the saucepan and gently simmer for about 10 to 15 minutes, or until the meatballs are cooked through and the flavors are well blended. Serve over hot cooked rice.

Makes 4 to 6 servings.

CHICKEN AND SWEET POTATOES IN ORANGE SAUCE

(main dish)

The sweet potatoes cook right along with the poultry in this easy and flavorful main dish.

1	pound boned and skinned chicken breast meat (from about 4 medium-sized breast halves), OR 1 pound boned raw turkey breast cutlets
2½	tablespoons enriched all-purpose or unbleached white flour
2	tablespoons butter or margarine
1½	cups orange juice
¼	teaspoon ground cinnamon
⅛	teaspoon ground ginger
	Pinch of black pepper, preferably freshly ground
2	medium-sized sweet potatoes, peeled and cut into ½-inch cubes (about 2 cups)
½	cup raisins or dried currants

To Serve
Hot cooked white or brown rice

Cut the chicken or turkey breast cutlets into small pieces, about 1 to 1½ inches square, and toss them with the flour. In a large deep skillet, melt the butter over medium-high heat; then lightly brown the poultry on both sides.

Mix the orange juice with the cinnamon, ginger, and pepper and add the mixture to the skillet along with the sweet potatoes and raisins. Bring to a boil; then cover and lower the heat. Simmer, stirring occasionally, for about 30 to 35 minutes, or until the poultry and sweet potatoes are tender and the sauce is thickened. (If the sauce becomes too thick, stir in a few teaspoons of water. If it's too thin, uncover the skillet and simmer the mixture for about 5 minutes longer, or until some liquid has evaporated and the sauce has thickened.) Serve over hot cooked rice.

Makes about 4 servings.

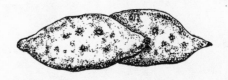

SWEET POTATO MUFFINS

1   cup enriched all-purpose or unbleached white flour
½   cup whole wheat flour
⅓   cup sugar
¼   cup instant nonfat dry milk powder
2   teaspoons baking powder
½   teaspoon ground cinnamon
¼   teaspoon ground cloves
⅛   teaspoon ground nutmeg
1   large egg, lightly beaten
3   tablespoons vegetable oil
¾   cup water
½   cup (packed) mashed cooked or canned sweet potatoes
½   cup raisins or dried currants

In a medium-sized bowl, combine the flours, sugar, milk powder, baking powder, and spices and mix well.

In a small bowl, combine the egg, oil, water, and sweet potatoes. Stir with a fork to mix well.

With a large spoon, stir the sweet potato mixture into the dry ingredients. Stir just until mixed. Fold in the raisins. Spoon into 12 medium-sized, lightly greased or nonstick spray-coated muffin cups. Bake the muffins in a preheated 400-degree oven for about 16 to 19 minutes, or until they are lightly browned. Serve warm.

Makes 12 muffins.

SOUTHERN-STYLE SWEET POTATO PIE

(dessert)

This is an old favorite with new twists—orange juice and honey—to give it a unique flavor. The crust is prebaked (and, if desired, coated with egg white) to help keep the bottom crisp after it is filled. To save time, bake the pastry shell while the sweet potatoes are cooking.

1   unbaked pastry shell for a 9-inch pie
    About 1 to 2 teaspoons egg white, lightly beaten, optional (may be taken from the egg whites in the filling)

Filling
½   cup orange juice
2   large eggs

2   large egg whites
2   cups mashed, cooked sweet potatoes (from about 3 medium-sized ones
    weighing a total of about 2 pounds)*
½   cup sugar
2   tablespoons honey
1   tablespoon cornstarch
1   teaspoon vanilla extract
1   teaspoon ground cinnamon
½   teaspoon ground nutmeg
¼   teaspoon ground allspice

Use a fork to prick holes all over the pastry shell. Press a large piece of aluminum foil on top of the crust; then fill the bottom with any kind of dry beans or some aluminum "pie weights." (This helps prevent crust shrinkage during baking. The beans or pie weights can be reserved, and used repeatedly for the same purpose.) Bake the shell in a preheated 400-degree oven for 15 minutes. Use the foil to lift out the beans and set both aside. Return the crust to the oven and bake it for an additional 7 to 10 minutes, or until it is very lightly browned. (To help ensure crispness, brush the inside of the shell lightly with the egg white and return it to the oven for 1 to 2 minutes longer, or until the white is set.) Cool the shell on a wire rack until it is needed.

For the filling, put all the ingredients into a food processor or blender in the order listed and process until completely combined and smooth. Pour the filling into the prebaked shell and bake in a preheated 350-degree oven for about 55 to 60 minutes, or until the filling is set, and a knife inserted in the center comes out almost clean. Cool the pie on a rack before serving. Cover and refrigerate for longer storage.

Makes about 8 servings.

*Any method can be used to cook the sweet potatoes. A quick one is to peel them and cut them into ½- to ¾-inch cubes. There should be about 4 cups. Put the cubes in a saucepan with about 1 inch of water. Simmer them, covered, until very tender, about 15 to 20 minutes; then drain them immediately and mash them while they are still warm.

Tomatoes

Although tomatoes are now the most popular vegetable for home gardeners, and third among canned vegetables, they were once shunned because they were thought to be poisonous.

A native of the Peruvian Andes Mountains, the original tomato plant had tiny yellow fruit we'd hardly recognize today. Spanish explorers introduced the tomato to the Old World during the sixteenth century. However, early European botanists recognized that tomatoes—like potatoes and eggplant—are related to belladonna (deadly nightshade). And, indeed, the leaves and stems of the plant do contain toxic substances. Thus, for two hundred years, tomatoes were grown in Europe mainly for ornamental purposes.

Ironically, those Europeans who were brave enough to eat the tomato thought it to be an aphrodisiac, and thus gave it the poetic moniker *pomme d'amour* or "love apple." It was known by that term for many years both in France and this country. (The name "tomato" derives from the old South American Indian word *tomatl.*)

In the eighteenth century, the Italians led the way in cultivating tomatoes specifically for use as food. Through extensive propagation, they developed a wide variety of luscious, large red globes and pear shapes, which eventually became an indispensable part of Italian cuisine. (According to some experts, the Italian name for tomato, *pomodoro*, comes from *pomo d'oro* which means "apple of gold" and harks back to the original tomatoes. Interestingly, there are still a few varieties of golden yellow and orange-yellow tomatoes in existence today.)

Thomas Jefferson was probably one of the first Americans to grow tomatoes for food. Eventually, other home gardeners followed his lead. But, due to the tomato's perishability and the mistaken belief that it had to be cooked for hours to remove any possible toxins, tomatoes did not become popular here until after the First World War.

The question is sometimes asked whether the tomato is a fruit or vegetable. Botanically, it is considered a berry because it is pulpy with soft, edible seeds. However, in an 1893 decision concerning import duties, the Supreme Court ruled that the tomato should be designated a vegetable. So, for trade purposes, it is legally a vegetable.

There are three main types of fresh tomatoes commonly available. These are the large, roundish or elliptical tomato such as the "beefsteak"; the pear-shaped "plum tomato" or "Italian tomato"; and the tiny, round "cherry tomato" (which is probably most similar to the tomato's wild ancestor). Although all come in both red and yellow varieties, the latter color is relatively rare.

Availability: While tomatoes can now be purchased the year round, the best ones with the most flavor are those grown locally during the summer and early fall. In some areas, plum tomatoes are available only seasonally.

Choosing the Best: Tomatoes should have firm, smooth skin and be well shaped and heavy for their size. Choose those that are intensely colored and have a good tomato smell. Avoid any with green or yellow patches at the end, or deep growth cracks, and also those that are very soft or have any soft spots. The very best tomatoes are locally grown ones that have been allowed to ripen completely and color on the vine.

Tomatoes supposedly reach maturity several days before they turn red, and those meant for long-distance shipping are often picked at this stage because very firm, green tomatoes hold up better than ripe ones. A few are allowed to slowly ripen naturally; most are "pushed" into "ripening" with ethylene gas. Even though they eventually turn red, these tomatoes will never have the rich flavor, juiciness, and velvety texture of real vine-ripened ones, but instead will tend to be mealy, dry, and tasteless.

Be aware that some tomatoes sold as "vine-ripened" have only just begun to turn pink when picked, and are not as tasty as those that have completely ripened and colored on the vine.

Nutritional Value: Tomatoes are an excellent source of vitamin C and are also high in vitamin A, potassium, phosphorus, and other minerals.

Storage: Tomatoes that are not completely red should be left at room temperature to ripen. To hurry the process, put tomatoes in a paper bag with a very ripe apple (which gives off small amounts of ethylene gas). Refrigerate red tomatoes. Seasonal, vine-ripened tomatoes will keep about 2 to 3 days in the refrigerator. Firm, "gassed," winter tomatoes will keep up to 2 weeks in the refrigerator.

Preparation and Basic Cooking: Fresh, large tomatoes need only be rinsed and cored before serving them raw. They can be cut into slices (some say vertical slices leak less than horizontal ones) or into wedges. Cherry tomatoes need not be cored; simply remove the stem if it is still attached. The smaller ones can be served whole; the larger ones, halved.

Some recipes call for peeled and/or seeded tomatoes. To quickly peel whole tomatoes, blanch them in boiling water for about 30 seconds; then plunge them into cold water to cool slightly. The skin should come off easily. To seed tomatoes, cut them in half vertically; then squeeze gently, so that the seeds fall out. Or, if you prefer, scrape the seeds out with a spoon.

Simple Serving Suggestions: Raw tomatoes are naturals in tossed salads and sandwiches. Home-grown tomatoes are quite good with an herbed vinaigrette dressing. Raw tomatoes can also be stuffed with a variety of salads, such as shrimp, tuna, or egg salad. Cut the tomato into wedges that stay attached at the bottom, open the tomato like a flower, and put the salad in the center. Or scoop the flesh and seeds out of the tomato to make a shell and fill the shell with salad. Tomato shells can also be stuffed with a meat or vegetarian filling and then baked.

Whole small green tomatoes can be pickled like cucumbers.

CHERRY TOMATOES STUFFED WITH CAULIFLOWER-CHEESE FILLING

(hors d'oeuvre or side dish)

Cauliflower gives the filling an intriguing, interesting flavor that complements the seasonings quite well. It's a great way to use up leftover cooked cauliflower.

Filling

1½	cups finely chopped *cooked* cauliflower
1	cup part-skim ricotta cheese (if unavailable, substitute regular ricotta)
¼	cup very finely chopped celery
2	tablespoons finely chopped fresh parsley leaves
1	tablespoon instant minced onions
1	tablespoon water
1	tablespoon dried chopped chives (or 1 tablespoon chopped fresh chives)
1	teaspoon dried basil leaves (or 1 tablespoon chopped fresh basil leaves)
½	teaspoon Worcestershire sauce
½	teaspoon salt (or to taste)
⅛	teaspoon black pepper, preferably freshly ground
3 to 4	drops Tabasco sauce

Tomatoes

1	pint cherry tomatoes, washed, dried, and stems removed

Put the cauliflower in a bowl and mash it with a fork until it is very coarsely puréed. (Do not use a food processor; it makes the cauliflower too smooth.) Stir in the ricotta, celery, and parsley. Soften the instant minced onions in the water; then stir them into the cauliflower mixture. Add the remaining filling ingredients and stir to combine well. Refrigerate the filling for several hours to allow the flavors to mingle. Stir it again before using and adjust the seasonings, if necessary.

To prepare the tomatoes, cut a slice from the end *opposite* the stem end of each tomato. (They are less likely to wobble if set on a tray with the stem end down.) Reserve the slices. Use a small spoon, preferably a serrated grapefruit spoon, to scoop out the seeds and pulp (reserve for another use or discard) from each tomato. Turn the tomato shells upside down onto some paper towels, and allow them to drain for about 5 minutes. Then use a small spoon or pastry bag (with a plain tip) to stuff each tomato with the filling so that it is slightly heaped on top. Loosely set a reserved slice on top of the filling, tilting it to one side, as a "lid" for each tomato.

The filling may be prepared up to 2 days in advance, and the tomatoes stuffed up to about 8 hours ahead of serving time.

Makes about 20 to 30 hors d'oeuvres (depending on the size of the tomatoes) or 4 to 6 side-dish servings.

MOLDED GAZPACHO SALAD

2	packets unflavored gelatin
2½	cups tomato juice, divided
1	beef bouillon cube
1	tablespoon red wine vinegar or apple cider vinegar
¼	teaspoon salt
¼	teaspoon black pepper, preferably freshly ground
1 to 2	drops Tabasco sauce
1½	cups chopped fresh tomatoes
1½	cups chopped cucumber
⅓	cup finely chopped red sweet pepper (if unavailable, substitute a sweet green pepper)
⅓	cup finely chopped celery
1	tablespoon finely chopped fresh chives (or ½ tablespoon dried chopped chives)
1½ to 2	cups torn crisp greens, such as endive or romaine lettuce

Sprinkle the gelatin over ½ cup of the tomato juice in a small saucepan. Set aside for 5 minutes, or until the gelatin softens. Stir the mixture and add the bouillon cube. Place the saucepan over medium-high heat and heat, stirring, until the gelatin and bouillon cube dissolve. Remove from the heat.

Stir together the heated juice mixture, the remaining 2 cups of tomato juice, and all the other ingredients until well blended. Transfer the mixture to a lightly oiled 1-quart mold. Cover and refrigerate for 3½ to 4 hours, or until the salad is set. Unmold the salad on a bed of greens and serve.

Makes 6 to 8 servings.

TOMATOES OREGANO

Prepare this colorful salad plate when vine-ripened tomatoes are in season. It can be made ahead, looks great on a buffet table, and always wins compliments.

6 to 7   large vine-ripened tomatoes

Marinade
- ¼   cup red wine vinegar or apple cider vinegar
- ¼   cup vegetable oil
- 2   tablespoons chopped fresh chives (or 1 tablespoon dried chopped chives)
- 1½   teaspoons dried oregano leaves
- 1½   teaspoons sugar
- ½   teaspoon salt
- ¼   teaspoon black pepper, preferably freshly ground

Garnish
Watercress or parsley sprigs

Bring about 3 cups of water to a boil in a small saucepan over high heat. One at a time, transfer the whole tomatoes to the pan with a slotted spoon and immerse them in the boiling water for 10 seconds. Immediately transfer the tomatoes to a colander and let them stand until cool enough to handle. Core the tomatoes and peel off the skins with a paring knife. Cut each tomato crosswise into about 4 thick slices. Attractively arrange the slices in one layer in a large shallow dish or serving plate.

Combine all the marinade ingredients in a small jar or cruet. Vigorously shake the contents until the sugar dissolves and the ingredients are well blended. Drizzle the marinade over the tomatoes, being sure to moisten each slice. Cover the tomatoes with plastic wrap and refrigerate for at least 1 hour and up to 8 hours. Just before serving, garnish the plate with watercress or parsley sprigs.

Makes 6 to 8 servings.

PASTA WITH FRESH TOMATO "SAUCE"

(side dish)

In this recipe, brief cooking retains the wonderful fresh tomato flavor and also some of the texture. The light "sauce" is thus quite different from traditional, long-cooked tomato sauces. (Note: If you dislike bits of tomato skin and seeds in the sauce, peel and seed the tomatoes before cooking.)

8   ounces pasta, such as spaghetti, linguini, or noodles
2   tablespoons good-quality olive oil
1   small onion, finely chopped
1   garlic clove, minced
4   large vine-ripened tomatoes, diced (first peeled and seeded, if desired)
1   medium-sized sweet green pepper, diced
2   tablespoons finely chopped fresh parsley leaves
1   teaspoon dried basil leaves (or 2 teaspoons finely chopped fresh basil)
½   teaspoon dried chervil or marjoram leaves (or 1 teaspoon finely chopped fresh chervil or marjoram)
½   teaspoon dried oregano leaves (or 1 teaspoon chopped fresh oregano)
¼   teaspoon salt
⅛   teaspoon black pepper, preferably freshly ground

Cook the pasta according to the package directions (omitting the salt, if desired). While the pasta is cooking, prepare the vegetable "sauce."

In a large skillet over medium-high heat, heat the oil; then cook the onion and garlic, stirring, until they are tender but not browned. Add the remaining ingredients and cook, stirring, about 5 minutes, or until the vegetables are hot and just cooked through. Drain the pasta well; then toss it with the vegetables.

Makes about 4 servings.

BRAISED TOMATOES WITH HERBS

(side dish)

Here is a tasty way to use up an overabundance of vine-ripened tomatoes. In this dish, chopped tomatoes are used to make a sauce for whole ones.

2	tablespoons butter or margarine
¼	cup chopped onion
⅓	cup chopped sweet red pepper
2	cups peeled and chopped vine-ripened tomatoes (about 2 medium-sized ones)
1½	tablespoons chopped fresh chives (or 1½ teaspoons dried chopped chives), divided
⅛	teaspoon dried basil leaves
¾	teaspoon salt
⅛	teaspoon black pepper, preferably freshly ground
	Pinch of cayenne pepper
4 or 5	large vine-ripened tomatoes (about ½ pound each), cored and peeled (but left whole)

Over medium-high heat, melt the butter in a saucepan or pot large enough to hold the whole tomatoes. Add the onion and sweet pepper and cook, stirring, for 3 to 4 minutes, or until the onion is limp. Add the *chopped* tomatoes and cook, stirring occasionally, for 4 to 5 minutes, or until the tomatoes are soft and some of the juice has evaporated. Transfer the mixture to a blender and blend on high speed for about 1 minute, or until completely puréed. Return the mixture to the saucepan. Stir about half the chives, the basil, salt, black pepper, and cayenne pepper into the purée. Add the whole tomatoes to the pan, spooning some of the puréed mixture over each. Bring the puréed mixture to a simmer and cook the tomatoes, uncovered, for 14 to 18 minutes, or until they are tender when pierced with a fork, but still hold their shape.

With a slotted spoon, transfer the tomatoes to individual serving bowls. Raise the heat to high and boil the purée for 1 minute, or until it is slightly thickened. Spoon the puréed mixture over the tomatoes, dividing equally among them. Garnish the tomatoes with the remaining half of the chives and serve.

Makes 4 to 5 servings.

ASPARAGUS-STUFFED TOMATOES

(side dish)

Eye-catching and elegant, this makes a nice addition to a festive meal.

5 to 6	large vine-ripened tomatoes, cored
1	10-package frozen asparagus spears, thawed and drained
2	tablespoons butter or margarine
1	medium-sized onion, finely chopped
¼	teaspoon salt
⅛	teaspoon celery salt
⅛	teaspoon black pepper, preferably freshly ground
⅓	cup fresh bread crumbs
⅓	cup coarsely grated or shredded sharp Cheddar cheese

Carefully scoop out the pulp and seeds from the tomatoes, leaving a ¼-inch-thick "wall" in each. Reserve the pulp and seeds in a small bowl. Invert the tomato shells on a plate and let them drain while the other ingredients are being readied.

Trim the tips from the asparagus spears and reserve. Cut the spears into ¼-inch-long pieces, discarding any tough ends; reserve the pieces separately.

Melt the butter in a medium-sized skillet over medium-high heat. Add the onion and cook, stirring for 3 minutes, or until it is limp. Add the reserved tomato pulp and seeds, along with the salt, celery salt, and pepper. Cook, stirring, for 4 to 5 minutes longer, or until the mixture thickens slightly. Stir in the reserved asparagus pieces and cook, stirring for 2 to 3 minutes longer, or until the asparagus is just tender. Remove the pan from the heat and stir in the bread crumbs and cheese until the mixture is well blended.

Using a teaspoon, stuff the tomatoes with the asparagus-tomato mixture, dividing it equally among them. Arrange the tomatoes upright in a baking dish or pie plate. Bake them, uncovered, in a preheated 350-degree oven for 14 to 17 minutes, or until they are heated through but still hold their shape.

Meanwhile, combine the reserved asparagus tips in a small saucepan with a small amount of water. Simmer the tips for 2 to 3 minutes, or until they are just tender. When the tomatoes are done, arrange a few asparagus tips on the top of each and serve.

Makes 4 to 5 servings.

TACO SALAD

(main dish)

Colorful, healthful, and always welcomed by children and adults alike.

Filling

1   tablespoon vegetable oil
1   small garlic clove, minced
1   large onion, chopped
1   medium-sized celery stalk, chopped
½   cup finely chopped sweet green pepper
½   cup finely chopped fresh parsley leaves
1   large tomato, chopped
2   15- to 16-ounce cans red kidney beans, well drained and coarsely mashed (using a fork or the bottom of a glass)
1   tablespoon chili powder
¼   teaspoon salt
¼   teaspoon celery salt

Garnishes and Toppings

1   4½-ounce package tostada shells, broken into bite-sized pieces
2   cups shredded iceberg, romaine, or other crisp lettuce
6   ounces mild Cheddar or Longhorn cheese, shredded or grated (about 1½ cups packed)
4   medium-sized tomatoes, chopped
2   teaspoons apple cider vinegar
¼   teaspoon salt
½   cup chopped scallions, including green tops (optional)
1   small avocado, peeled, seeded, and diced
¼   cup pitted halved ripe (black) olives (optional)

Prepare the filling as follows: Combine the oil, garlic, and onion in a large skillet over medium-high heat and cook the vegetables, stirring, for 2 to 3 minutes. Add the celery, green pepper, and parsley and cook, stirring occasionally, for 4 to 5 minutes longer, or until the vegetables are limp. Stir in the chopped tomato, kidney beans, chili powder, salt, and celery salt. Lower the heat to medium and cook the mixture for 4 to 6 minutes, or until it is well blended and slightly thick. Remove the filling from the heat and set aside to cool slightly.

Assemble the salads on 4 individual serving plates as follows: Spread about three fourths of the tostada shell pieces on the plates, dividing them equally among them. Srpinkle half the shredded lettuce over the tostada pieces. Mound the bean filling in the center of the plates, dividing the mixture equally among them. Top the filling with the shredded cheese, then with the remaining lettuce and tostada shell pieces.

Stir the chopped tomato, vinegar, and salt together in a medium-sized bowl until well mixed. Spoon the seasoned tomatoes over the cheese and lettuce. Top the salads with the chopped scallions (if desired) and avocado. Garnish the plates with the halved olives and serve immediately.

Makes 4 servings.

EGGS BAKED IN TOMATO SHELLS

(light main dish or side dish)

This dish looks very pretty and is perfect for a brunch. It's worth saving for gorgeous seasonal tomatoes, rather than the "winterized" pale red "tennis balls" available in most supermarkets at other times in the year.

4 large, firm, vine-ripened tomatoes
   About ⅛ teaspoon salt
½ cup fresh whole wheat bread crumbs (made in a food processor or blender from about 1 slice of bread)
1 teaspoon dried basil leaves (or 1 tablespoon finely chopped fresh basil leaves)
1 teaspoon dried oregano leaves (or 1 tablespoon finely chopped fresh oregano leaves)
¼ cup grated Parmesan cheese
⅛ teaspoon black pepper, preferably freshly ground
4 large eggs

Cut a thin slice from the stem end of each tomato. Scoop out the seeds and membranes and reserve for another use or discard. Lightly sprinkle the inside of each tomato with salt; then invert the tomatoes on paper towels and let them drain for about 10 minutes.

Meanwhile, stir together the bread crumbs, basil, oregano, cheese, and pepper. Spoon 2 tablespoons of the mixture into the bottom of each tomato; then place each tomato, upright, in an ovenproof custard cup. Carefully crack open an eggshell, and gently let the raw egg slip into one of the tomatoes so that the yolk does not break open. Repeat for the remaining 3 eggs and tomatoes. Sprinkle the remaining crumb mixture on top of the eggs, using about 1 tablespoon for each.

For ease in handling, place the custard cups inside a baking pan, and bake the tomatoes in a preheated 375-degree oven for about 15 to 20 minutes, the total time depending on the firmness desired in the cooked egg. Serve hot.

Makes 4 servings.

TOMATO-CHEESE PIZZA

(main dish)

Dough

3½ to 4	cups enriched all-purpose or unbleached white flour
1	packet fast-rising dry yeast
1	teaspoon sugar
¾	teaspoon salt
1¼	cups hot water (125 to 130 degrees)
1	tablespoon olive or vegetable oil

Sauce

1½	tablespoons olive oil
1	small garlic clove, minced
1	medium-sized onion, finely chopped
½	cup finely chopped fresh parsley leaves
⅓	cup finely chopped sweet red pepper (if unavailable, substitute sweet green pepper)
1¼	cups canned tomato sauce
1	6-ounce can tomato paste
1½	teaspoons dried oregano leaves
1	teaspoon dried marjoram leaves
½	teaspoon sugar
⅛	teaspoon black pepper, preferably freshly ground
	Pinch of crushed hot red pepper

Toppings

12	ounces mozzarella cheese, grated (3 cups packed)
¼	cup grated Parmesan cheese
1½	cups sliced fresh mushrooms
2	tablespoons olive oil
1	small sweet red or green pepper, cut into thin rings
½	cup chopped scallions, including green tops (optional)
½	cup sliced pitted ripe (black) olives (optional)

To prepare the dough stir together ¾ cup of the flour, yeast, sugar, and salt in a cup or small bowl until well blended.

Pour the water into a blender. With the blender on low speed add the flour mixture and blend for 15 seconds. Transfer the mixture to a medium-sized bowl. Vigorously stir in 2¼ cups more flour until blended. Working in the bowl, gradually knead in about ½ to 1 cup more flour, or enough to yield a smooth and malleable but not stiff dough. Transfer the dough to a greased medium-sized bowl. Cover the bowl with plastic wrap and set aside in a very warm place for 25 minutes, or until doubled in bulk.

Meanwhile, prepare the sauce: Combine the oil, garlic, and onion in a medium-sized saucepan over medium-high heat. Cook, stirring, until the onion is limp. Stir in all the remaining sauce ingredients and bring the mixture to a boil. Lower the heat and simmer, uncovered, for 10 minutes. Set the sauce aside.

Prepare the topping ingredients and set them aside.

Punch down the risen dough with lightly greased hands. Divide the dough in half. Put each half of the dough into an oiled 12-inch-diameter pizza pan. Press the dough in an evenly thick layer out to the pan edges. Press up over the pan edges about ¼ inch to form the edge of the crust. Spread the sauce over the dough, dividing equally between the two pans. Top the sauce with the cheeses. Combine the mushrooms and olive oil, stirring until the mushrooms are coated with the oil. Sprinkle the mushrooms over the pizzas. Then sprinkle them with the remaining topping ingredients.

Bake the pizzas in a preheated 475-degree oven for 18 to 23 minutes, or until the tops are well browned and bubbly. Cut the pizzas into wedges and serve.

Makes 2 pizzas or 6 to 7 servings.

Turnips

The bulbous, edible tap roots known as turnips have been used for food since ancient times. In the first century A.D., the Roman scholar Pliny mentioned several long, flat, and round turnip varieties called "rapa" and "napus." Later sources noted that the napus variety was growing in the sand along the sea coasts of Holland, Sweden, and England.

At some point, the napus turnip also became known to the Anglo-Saxons, who shortened the Latin name to "neap." This word was eventually modified and embellished with the prefix "turn"—meaning to make round. The resulting compound word evolved into our modern English designation, "turnip."

Although the turnips available in modern supermarkets usually weigh only a few ounces apiece when harvested, many of the early references describe specimens of truly mammoth proportions. A sixteenth-century writer noted that he had heard of turnips weighing 100 pounds, and had actually seen some long, purple ones weighing about 30. An eighteenth-century source wrote that the greatest weight he was aware of was 36 pounds.

A member of the cabbage family, the turnip was introduced into the New World by Cartier when he visited Canada in 1540. The vegetable flourished there and quickly spread southward. It was being cultivated in the American colonies by the 1600s.

Although turnips are a standard offering in produce sections during the fall and winter, they tend to be overlooked by the American consumer. They may be eaten raw, but are usually served cooked, most often with other vegetables in soups, stews, and boiled dinners. Many Southern cooks also make use of the leafy tops of the plant, boiling them along with seasoning meat (such as bacon and salt pork) and serving as "greens."

Availability: Turnips are most abundant from September through March. The supply dwindles to almost nothing in midsummer. Particularly in the South, fresh turnip leaves, or greens, are available in some supermarkets during the fall and winter months. Depending on consumer demand, frozen and canned turnip greens are also stocked in some stores year round.

Choosing the Best: Select turnips that are firm, well shaped, and free from cuts, blemishes, and growth cracks. Also, choose the smaller turnips (3 to 3½ inches in diameter or less), as these are more tender and have a sweeter flavor. When turnips are sold with the tops are still intact, these should be green and springy; limpness and yellowing indicate deterioration.

Most turnip greens are sold prepackaged. However, if buying loose greens, choose leaves that are crisp and clear green.

Nutritional Value: Turnips are a good source of vitamin C. Turnip greens are also an excellent source of vitamin A.

Storage: Store turnips in a very cool, dry pantry for up to 1 week, or in the refrigerator in a plastic bag for up to 2 to 3 weeks. Turnips greens should be washed and thoroughly drained, and then loosely packed in plastic bags and refrigerated. They are quite perishable and should be eaten in a day or two.

Basic Preparation and Cooking: Wash the turnips and peel with a sharp knife or vegetable peeler. (Small, tender turnips may be left unpeeled, if desired.) If the vegetables were purchased with tops, also cut off the leaves level with the stem caps or crowns. The flesh can then be sliced, diced, or shredded and served raw.

To prepare turnip greens, wash well and drain; then rinse again. Trim off any tough stems or midribs using a sharp knife. Tear the leaves into large bite-sized pieces.

To cook turnips, simmer 1-inch cubes in a small amount of water or bouillon until they are just tender, about 10 to 15 minutes. Or steam for 20 to 25 minutes. Boil turnip greens in water or beef bouillon for 12 to 15 minutes, or until they are just tender. Or steam them for about 20 minutes.

Simple Serving Suggestions: Add diced or shredded raw turnips to salads. Dress cooked turnips with butter, salt, and pepper and perhaps a pinch of dillweed or tarragon. Dress turnip greens with butter, salt, pepper, and a dash of lemon juice.

PEASANT SOUP

1   tablespoon butter or margarine
1   large onion, finely chopped
1   garlic clove, minced
1½ cups grated or shredded peeled potato (about 1 medium-sized potato)
1   medium-sized carrot, grated or shredded
1   medium-sized turnip, peeled and diced
3   cups water
2   cups chicken stock or bouillon (reconstituted from cubes or granules)
1   8-ounce can tomato sauce
1   bay leaf
½   teaspoon celery salt
½   teaspoon dried basil leaves
½   teaspoon dried marjoram leaves
½   teaspoon sugar
¼   teaspoon chili powder
¼   teaspoon powdered mustard
⅛   teaspoon dried thyme leaves
¼   teaspoon salt
¼   teaspoon black pepper, preferably freshly ground

In a large saucepan or small Dutch oven, combine the butter, onion, and garlic over medium-high heat. Cook the onion and garlic, stirring, until the onion is tender. Add all the remaining ingredients. Bring to a boil. Cover and lower the heat. Simmer for 20 to 25 minutes, or until the vegetables are tender and the potato has thickened the soup slightly.

Makes 5 to 6 servings.

TURNIP AND CARROT SLAW

(side dish)

Adding raw turnip to slaw might seem unusual, but it complements the more typical ingredients, such as carrots and cabbage, very well. In fact, the turnip adds a very pleasant, yet subtle, radish-like zest.

2 medium-sized turnips, peeled and shredded or coarsely grated
1 large carrot, shredded or coarsely grated
1 tablespoon finely chopped red onion
3½ cups shredded green cabbage

Dressing
⅓ cup mayonnaise
⅓ cup commercial sour cream
3 tablespoons apple cider vinegar
2 teaspoons sugar
¼ teaspoon celery salt
¼ teaspoon powdered mustard
¼ teaspoon dried dillweed
⅛ teaspoon onion salt
⅛ teaspoon black pepper, preferably freshly ground

Combine the turnips, carrot, red onion, and cabbage in a large salad bowl and toss until well mixed.

Combine all the dressing ingredients in a small deep bowl. Beat the mixture with a wire whisk or a fork until it is completely blended and smooth. Add the dressing to the vegetables and toss until well blended. Cover and refrigerate the slaw for at least 30 minutes and up to 24 hours before serving.

Makes 5 to 7 servings.

CHEESY TURNIP-RUTABAGA PUFF

(side dish)

Smooth and rich tasting, this dish is a great way to introduce your family to turnips and rutabagas—two tasty, healthful vegetables, which are often relegated just to soups and stews.

4   cups peeled and diced turnips (about 3 medium-sized ones)
2   cups peeled and diced rutabagas (about 1 large one)
2   tablespoons butter or margarine
1   small onion, finely chopped
3   tablespoons enriched all-purpose or unbleached white flour
¼   teaspoon dried basil leaves
¼   teaspoon salt
⅛   teaspoon black pepper, preferably freshly ground
2   large eggs, separated
⅓   cup instant nonfat dry milk powder
¼   cup plus 1 tablespoon grated Parmesan cheese

Put the turnips and rutabagas in a 2½-quart saucepan with about 1 inch of water. Bring to a boil; then lower the heat and simmer, covered, for about 15 to 20 minutes, or until the vegetables are very tender. Drain well. Purée the cooked vegetables in a food processor or food mill. This should yield about 3 cups of purée. Set aside. Wash and dry the saucepan.

In the saucepan, over medium-high heat, melt the butter; then cook the onion until it is tender but not browned. Stir in the flour until combined and cook, stirring, for about 30 seconds. Remove from the heat. To the onion mixture in the saucepan, add the vegetable purée, basil, salt, pepper, egg yolks, milk powder, and ¼ cup of the cheese. Stir until combined.

In a clean mixing bowl, beat the egg whites just until stiff peaks form. Fold the whites into the vegetable mixture. Turn out into a lightly greased or nonstick spray-coated 10-inch-diameter quiche pan or 9-inch-square oven-to-table casserole (or equivalent). Use a spatula to form some decorative swirls and small peaks in the top of the vegetable mixture. Sprinkle the remaining 1 tablespoon of cheese on top.

Bake in a preheated 350-degree oven for about 20 to 25 minutes, or until set. To serve, spoon directly from the casserole.

Makes 6 to 8 servings.

BRAISED TURNIPS

(side dish)

Although this is rather simple to prepare, it makes a very tasty and elegant accompaniment to hearty roasts or chops.

2 teaspoons sugar
¾ cup (plus more, if needed) beef broth or bouillon (reconstituted from cubes or granules)
1 tablespoon butter or margarine
½ teaspoon lemon juice, preferably fresh
⅛ teaspoon black pepper, preferably freshly ground
   Generous 1 pound turnips (about 5 medium sized), peeled and cut crosswise into ⅛-inch-thick slices
   About 1 tablespoon finely chopped fresh parsley leaves for garnish (optional)

Heat the sugar in a large heavy skillet over high heat until the crystals just begin to melt and turn golden. Immediately add about ½ cup of the beef broth to the skillet; the liquid will bubble and foam up the sides. When the bubbling subsides, lower the heat to medium and stir in the remaining ¼ cup of beef broth, butter, lemon juice, and pepper. Add the turnips and allow the mixture to come to a simmer. Adjust the heat if necessary and gently simmer the turnips for 12 to 15 minutes, or until they are almost tender. (Add a few more tablespoons of beef broth if needed to prevent the mixture from boiling dry.) When the turnips are just barely tender, raise the heat slightly and cook, stirring, until all the excess moisture has evaporated from the skillet. Transfer the turnips to a serving bowl and sprinkle with the chopped parsley leaves, if desired.

Makes about 4 servings.

TRICOLOR LAYERED VEGETABLE LOAF

(side dish)

Puréed carrots, turnips, and peas give the layers of this impressive loaf their glorious bright orange, white, and green colors. It can be assembled ahead of time and then baked shortly before serving, making it perfect for Thanksgiving and dinner parties.

(Note: A food processor makes this recipe very quick and easy. If it is not available, use a food mill to purée each vegetable; then add the remaining ingredients for each layer and mix well.)

Carrot Layer
- 1 pound carrots, scrubbed but not peeled, and cut into 1-inch-long pieces
- 1 tablespoon butter or margarine
- 1 large egg
- 2 tablespoons skim, lowfat, or whole milk
- 2½ tablespoons enriched all-purpose or unbleached white flour
- ¼ teaspoon salt
- ⅛ teaspoon ground ginger
- ⅛ teaspoon ground mace
- Pinch of freshly ground black pepper

Turnip Layer
- 1¼ pounds turnips (about 5 medium sized), peeled and cut into chunks (about 4½ cups)
- 1 tablespoon butter or margarine
- 1 large egg
- 1½ tablespoons skim, lowfat, or whole milk
- 2½ tablespoons enriched all-purpose or unbleached white flour
- ¼ teaspoon salt
- ⅛ teaspoon black pepper, preferably freshly ground

Pea Layer
- 3½ cups loose-pack frozen peas
- 1 tablespoon butter or margarine
- 1 large egg
- 3 tablespoons enriched all-purpose or unbleached white flour
- ¼ teaspoon salt
- Pinch of black pepper, preferably freshly ground
- Pinch of ground nutmeg

First, cook the carrots, turnips, and peas: Put each type of vegetable into a separate saucepan with water to cover. Bring to a boil; then lower the heat and simmer, covered, until the vegetables are tender. (The carrots and turnips will take about 20 minutes each, depending on the size of the pieces. The peas need only about 5 minutes.) Drain each vegetable well and cool it slightly.

While the vegetables are cooking, grease well or coat with nonstick spray an 8-by 4-inch loaf pan. Then line the bottom with a rectangle of wax paper and grease or spray the paper. Set aside.

For the carrot layer, put the cooked, drained carrots into the bowl of a food processor with the butter and finely chop them. Add the remaining carrot-layer ingredients, and process until the carrots are completely puréed. Stop the processor once or twice and scrape down the sides of the bowl with a rubber scraper to make sure all the carrots are processed. Spread the carrot purée evenly in the bottom of the prepared loaf pan. Rinse and dry the food processor bowl and blade.

Repeat the procedure with the turnip-layer ingredients, and spread the turnip purée evenly over the carrots. Repeat it again with the pea-layer ingredients, spreading the pea purée over the turnips. Cut a rectangle of wax paper to fit inside the pan and completely cover the pea mixture; grease or spray it, and place it, greased-side down, in the pan against the peas. (At this point, the prepared mold may be refrigerated for several hours or overnight, if desired.)

Bake the loaf (with the wax paper on top) in a preheated 350-degree oven for about 40 to 50 minutes, or until the top of the loaf has begun to shrink from the sides of the pan. (The loaf will need the longer baking time if it has been refrigerated.) Remove the pan from the oven. Cool the loaf in the pan for about 5 minutes. Carefully peel the wax paper from the top. Run a knife around the edge of the loaf to loosen it; then unmold it onto a large serving platter, and peel off the remaining piece of wax paper. Serve the loaf warm or at room temperature. Cut it crosswise into 1-inch-thick slices and use a wide cake server or large metal pancake turner to serve it.

Makes about 8 servings.

TURNIPS STUFFED WITH CHEESE

(side dish)

Here's an unusual and appealing, yet very easy, way to serve this nutritious vegetable.

4 medium-sized turnips, peeled but left whole
2 ounces sharp Cheddar cheese (or your choice), shredded (½ cup packed)
½ cup whole wheat bread crumbs (made in a food processor or blender from about 1 slice fresh or slightly stale whole wheat bread)
2 tablespoons butter or margarine, melted

Cook the turnips in boiling water until they are very tender (about 30 minutes, but the time depends on their exact size); then drain them well. Let the turnips cool until they can be handled. Use a metal melon baller, serrated grapefruit spoon, or small sharp knife to scoop out the centers, leaving a ⅜-inch-thick shell. (Reserve the scooped-out flesh.) Gently place the shells (they will be fragile) in a small baking dish.

Finely chop the reserved turnip flesh; then mix it with the cheese, bread crumbs, and melted butter. Carefully stuff each turnip with the mixture, piling it high on top. (This recipe may be prepared in advance to this point.)

Shortly before serving, bake the stuffed turnips, uncovered, in a 350-degree oven for about 25 minutes, or until the cheese is melted and bubbly.

Makes 4 servings.

BEEF AND VEGETABLE STEW

(main dish)

1   tablespoon vegetable oil
1   pound stew beef, cut into 1-inch cubes
1   large onion, finely chopped
1   garlic clove, minced
1   14- to 16-ounce can tomatoes, including juice
1   cup dry white wine
1   6-ounce can tomato paste
2   medium-sized carrots, cut into ¼-inch-thick slices
2   cups 2-inch fresh green bean pieces, stem ends removed
3   medium-sized turnips, peeled and cut into 1½-inch pieces
2   bay leaves
¾   teaspoon dried basil leaves
    Generous ¼ teaspoon dried thyme leaves
¾   teaspoon salt
¼   teaspoon black pepper, preferably freshly ground

Heat the vegetable oil in a large heavy pot over medium-high heat. In two batches, brown the meat in the oil. Return all the meat to the pot. Add the onion and garlic. Cook, stirring, until the onion is soft. Add the tomatoes, breaking them up with a spoon, along with the wine and tomato paste. Stir to make sure the tomato paste is well combined. Add all the remaining ingredients. Bring to a boil. Cover, lower the heat, and simmer for about 1½ hours, or until the vegetables and meat are tender. Stir occasionally to prevent sticking.

Makes 4 to 6 servings.

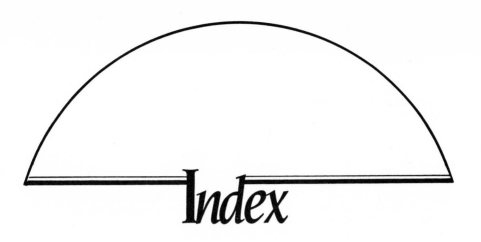

Index

turkey cutlets (or chicken breasts) with,
  sautéed, 132
Curry(ied)
  broccoli salad, 53
  cauliflower with peas, 105
  dip, 51
  potato-peanut salad, 212

Dill(ed)
  carrots, 91
  soup
    cauliflower-, 103
    kohlrabi, 152
Dip, curry, 51
Dried beans, *see* Beans, dried
Dumplings, chicken and parsley, 109–10
Dressing, 156
  calico coleslaw with cottage, 76
  for cauliflower-broccoli salad, 104
  creamy herb, Belgian endive with, 47
  cucumber, 130
  for curried potato-peanut salad, 212
  for Greek salad in pita pockets, 159
  for mixed green salad, 157
  zesty garlic, tossed salad with, 158

Eggplant(s), 133–38
  cheese-stuffed, rolls, 138
  Orientale, 137
  "oven-fried" breaded, 136
  preparation and basic cooking,
    134–35
  simple serving suggestions, 135
  and sweet pepper sauté, 135
  Turkish-style mixed vegetables, 171
Eggs
  asparagus-noodle bake, 16
  baked in tomato shells, 281
  carrot snack cake, 99
  cheesy turnip-rutabaga puff, 288
  Greek-style spinach "pie," 231
  quiche
    onion and tuna, 181
    scallion and cauliflower, 180
  salad Niçoise, 37–38
  spinach-cheese soufflé, no-fail, 230
Endive, *see* Belgian endive
English peas, *see* Peas
Escarole
  artichoke, avocado, and grapefruit
    salad, 6

Fava beans, 28
Fennel, 139–42
  about, 139–40
  with fresh tomato and garlic, 142
  with Parmesan cheese, braised, 141
  preparation and basic cooking, 140
  simple serving suggestions, 140
Feta cheese
  Greek salad in pita pockets, 159
  spinach "pie," Greek-style, 231–32
  stuffed Belgian endive, 46
Fish, "oven-fried," with salsa, 204–205
  see also Tuna; Salmon
Fresh beans, *see* Beans, fresh
Fruit casserole, microwave sweet potato and, 266

Garbanzo beans, *see* Chick-peas
Garlic, 173
  dressing, tossed salad with zesty, 158
  fennel with fresh tomato and, 142
Gazpacho salad, molded, 275
Gingered sweet potato casserole, 264
Grapefruit salad, artichoke, avocado and, 6
Great Northern beans, 18
  herbed white bean salad, 23
  and squash, Spanish-style, 258
Greek salad in pita pockets, 159
Greek-style spinach "pie," 231
Green beans, 27
  beef and vegetable
    soup, hearty, 223
    stew, 293
  chicken and fresh vegetable skillet dinner, 36
  hamburger skillet, tangy, 202
  Italian, and water chestnuts salad, 30
  marinated vegetables, 65
  minestrone, hearty meatless, 22
  with pizazz, 31
  salad Niçoise, 37–38
  with sesame seeds, 32
  shepherd's pie, 218
  Turkish-style mixed vegetables, 171
  vegetables in creamy tomato sauce, 33
Green chili pepper(s), 196–97
  chicken and, relleno casserole, 203
  corn soup, fiesta, 123
Green onions, 173, 174, 175
  see also Onion(s)
Green or white cabbage, 71
  see also Cabbage
Green pea soup, lettuce and, 56
Green pepper(s), sweet, 196–97
  cabbage and sweet peppers, braised, 77

Green pepper(s), sweet, (cont'd)
  chicken
    Chinese-style, and, 200–201
    Creole, 205
  fish with salsa, "oven-fried," 204–205
  hamburger skillet, tangy, 202
  marinated carrot and sweet pepper salad,
    89
  and onion sauté, 199
  pasta with, and onions, 199–200
  pepper steak, Chinese-style, 200–201
  and red pepper salad, roasted, 198
  squash and, skillet, 244
  stuffed, Mexican-style, 201
  veal skillet, Italian-style, 206
  vegetables
    and cheese, 107
    in creamy tomato sauce, 33
    Turkish-style, 171
Gruyère cheese for spinach-cheese soufflé, no-fail,
  230

Herbed broccoli with sesame seeds, 55
Herbed Brussels sprout soup, 64
Herbed cabbage and potato casserole, 78
Herbed mushroom soup, 162–63
Herbed white bean salad, 23
Horticultural beans, 28
Hubbard squash, 252
  see also Winter squash

Indian-style okra, 170–71
Indian-style peas and potatoes, 193–94
Italian green beans
  Turkish-style mixed vegetables, 171
  and water chestnuts salad, 30
Italian-style broccoli, 54–55
Italian-style veal skillet, 206

Kale, 143–49
  about, 143–44
  and cabbage salad, 146
  mixed vegetable stuffing or casserole, 148
  pork and vegetables one-pot dinner, savory,
    149
  and potato casserole, 147
  preparation and basic cooking, 144
  and red onion, 146–47
  simple serving suggestions, 144
  soup, velvety, 145
  spinach "pie," Greek-style, 231–32

Kidney beans, 17, 18
  burrito casserole, 25
  chicken and chilies relleno casserole, 203
  Mexican corn and, 124–25
  Mexican-style stuffed peppers, 201
  minestrone, hearty meatless, 22
  taco salad, 280
Kohlrabi, 150–53
  about, 150–51
  -dill soup, 152
  gratin, 153
  preparation and basic cooking, 151
  simple serving suggestions, 151

Large banana squash, 242
  see also Winter squash
Lasagne, spinach, 233–34
Leek(s), 173, 174, 175
  brown rice
    and barley bake with mushrooms, 179
    with carrots and, 177
  cauliflower vichyssoise, 102
  chicken with cabbage and noodles, 83
  onion and tuna "quiche," 181
  potato-, soup, 176
  see also Onion(s)
Lemon
  Brussels sprouts with, and Parmesan cheese, 69
  -glazed butternut squash cubes, 255
Lentils, 18
Lima beans, 18, 19, 28
  okra succotash, 169–70
  with sweet red pepper, 34
Lettuce, 154–59
  about, 154–56
  Greek salad in pita pockets, 159
  and green pea soup, 156
  mixed green salad, 157
  preparation, 155
  salad Niçoise, 37–38
  simple serving suggestions, 156
  taco salad, 280–81
  tossed salad with zesty garlic dressing, 158
Longhorn cheese
  taco salad, 280–81
  vegetables and, 107

Macaroni
  pasta-vegetable casserole, 244–45
  vegetable and ground beef skillet dinner, 81
  see also Pasta
Marinated artichoke hearts, 5

Marinated carrot and sweet pepper salad, 89
Marinated cucumber and cherry tomato salad, 130–31
Marinated vegetables, 65
Marinated zucchini and tomato salad, 243
Marrow beans, 19
Mexican corn and beans, 124–25
Mexican-style stuffed peppers, 201
Minestrone, hearty meatless, 22
Monterey Jack cheese for hearty "California" sandwich, 238–39
Mozzarella
  cheese-stuffed eggplant rolls, 138
  pasta-vegetable casserole, 244–45
  spinach lasagne, 233–34
  tomato-cheese pizza, 282–83
  zucchini
    -cheese casserole, 247
    stuffed with tuna and, 248
Muffins
  carrot-bran, 97
  microwave "corny," 125–26
  pumpkin, 260
  sweet potato, 269
Mushroom(s), 160–65
  about, 160–62
  basic preparation and cooking, 162
  confetti rice, 164
  crazy quilt brown rice, 93
  leek, brown rice, and barley bake with mushrooms, 179
  shrimp-stuffed, 165
  simple serving suggestions, 162
  slaw, 163
  soup, herbed, 162–63

Napa cabbage, 70, 71
Navy beans, 19
  beef bake with, and barley, 26
  black-eyed pea and bean soup, spicy, 21
  herbed white bean salad, 23
  and squash, Spanish-style, 258
Noodle(s)
  asparagus-, bake, 16
  chicken with cabbage and, 83
  see also Pasta

Okra, 166–71
  about, 166–68
  cooked with rice and tomatoes, 169
  Indian-style, 170–71
  preparation and basic cooking, 167–68

simple serving suggestions, 168
succotash, 169–70
and sweet red pepper stir-fry, 168
Turkish-style mixed vegetables, 171
Onion(s), 172–83
  about, 172–75
  braised, 176–77
  butternut squash with, and apples, 256
  cabbage with tarragon, 76–77
  chicken with artichokes and, 7
  leeks, see Leek(s)
  Oriental-style chicken, hot and spicy, 182–83
  in paprika sauce, 178
  pepper(s) and
    pasta with, 199–200
    sauté, 199
  preparation and basic cooking, 174–75
  red
    kale with, 146–47
    soup, 175
  rutabaga and, spicy, 224
  scallion and cauliflower quiche, 180
  simple serving suggestions, 175
  and tuna "quiche," 181
  vegetables and cheese, 107
  vegetable stock, 86
Oregano, tomatoes, 276
Oriental cucumber soup, 129
Oriental stir-fried steak, 240
Oriental-style chicken
  and bean sprouts, 239
  hot and spicy, 182–83
Oriental vegetable stir-fry, mixed, 192–93

Paprika sauce, onions in, 177
Parmesan cheese
  Brussels sprouts with lemon and, 69
  cheesy turnip-rutabaga puff, 288
  fennel with, braised, 141
  spaghetti squash with herbs and, 259
Parsley dumplings, chicken and, 109
Parsnips, 184–87
  about, 184–85
  glazed, 186
  preparation and basic cooking, 185
  serving suggestions, 185
  sherried, and carrots, 187
Pasta
  with fresh tomato "sauce," 277
  macaroni, see Macaroni
  noodles, see Noodle(s)
  with peppers and onions, 199–200